MEI STRUCTURED MATHEMATICS

SECOND EDITION

Statistics 4

Michael Davies

Series Editor: Roger Porkess

Hodder & Stoughton

A MEMBER OF THE HODDER HEADLINE GROUP

MEI Structured Mathematics

Mathematics is not only a beautiful and exciting subject in its own right but also one that underpins many other branches of learning. It is consequently fundamental to the success of a modern economy.

MEI Structured Mathematics is designed to increase substantially the number of people taking the subject post-GCSE, by making it accessible, interesting and relevant to a wide range of students.

It is a credit accumulation scheme based on 45 hour modules which may be taken individually or aggregated to give Advanced Subsidiary (AS) and Advanced GCE (A Level) qualifications in Mathematics, Further Mathematics and related subjects (like Statistics). The modules may also be used to obtain credit towards other types of qualification.

The course is examined by OCR (previously the Oxford and Cambridge Schools Examination Board) with examinations held in January and June each year.

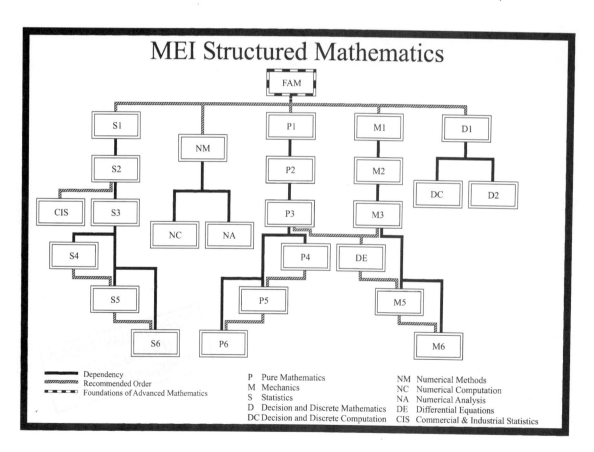

This is one of the series of books written to support the course. Its position within the whole scheme can be seen in the diagram above.

Mathematics in Education and Industry is a curriculum development body which aims to promote the links between Education and Industry in Mathematics at secondary level, and to produce relevant examination and teaching syllabuses and support material. Since its foundation in the 1960s, MEI has provided syllabuses for GCSE (or O Level), Additional Mathematics and A Level.

For more information about MEI Structured Mathematics or other syllabuses and materials, write to MEI Office, Albion House, Market Place, Westbury, Wiltshire, BA13 3DE.

Introduction

This is the fourth in a series of books written to support the Statistics modules in MEI Structured Mathematics.

By the time you are studying this module, you will already have come across and used all the fundamental ideas of statistics. This book has two aspects. In the chapters on modelling and estimation, you will re-examine and deepen your understanding of these fundamental concepts. In the remaining chapters you will build on what you already know by discovering how to test hypotheses and construct confidence intervals in new contexts. By the time you have finished this module, you will be acquainted with techniques which are used every day by working statisticians and be in a position to follow many of the statistical methods and arguments used in fields such as Archaeology, Psychology and Biology.

This is the second edition of *Statistics 4* in this series. This new edition includes a number of recent examination questions and in a few places the text has been rewritten to make it easier to follow. In addition this book also introduces a number of changes of notation in line with what has now become standard usage. Thanks are due to Michael Davies for his work in preparing the new edition.

Readers who are interested in a possible career involving Statistics may wish to consult the Royal Statistical Society's Career's web-site, www.rss.org.uk/careers for further information.

Roger Porkess

Contents

Contingency tables

Maryann was absent December 11–16 because she had a fever, sore throat, headache and upset stomach. Her sister was also sick, fever and sore throat, her brother had a low grade fever and ached all over. I wasn't the best either, sore throat and fever. There must be something going around, what do you think?

Parent's note to a teacher

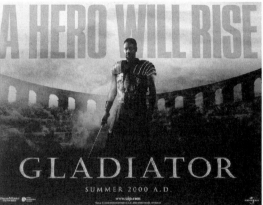

What kind of films do you enjoy?

To help it decide when to show trailers for future programmes, the management of a cinema asks its customers to fill out a brief questionnaire saying which type of film they enjoy. It wants to know whether there is any relationship between people's enjoyment of Horror films and Action movies. The management takes 150 randomly selected questionnaires and records whether those patrons enjoyed or did not enjoy Horror films and Action movies.

	Enjoyed Horror films	Did not enjoy Horror films
Enjoyed Action movies	51	41
Did not enjoy Action movies	15	43

This method of presenting data is called a 2×2 *contingency table* (later, you will meet larger contingency tables). It is used where two variables (here 'attitude to Horror films' and 'attitude to Action movies') have been measured on a sample, and each variable can take two different values (enjoy or not enjoy).

You have met *bivariate data* before, when you studied correlation in *Statistics 2*, but the difference is that there the values taken by the variables were numerical and took many (usually a continuous range of) values – you might have looked, for instance, at whether people's heights and weights are related. Here, the values taken by the variables fall into one or another of two categories, and are not numerical in nature.

It is conventional, and useful, to add the row and column totals in a contingency table: these are called the *marginal totals* of the table.

	Enjoyed Horror films	Did not enjoy Horror films	Marginal totals for Action movies
Enjoyed Action movies	51	41	92
Did not enjoy Action movies	15	43	58
Marginal totals for Horror films	66	84	150

At first sight, the data seem to support, for instance, the statement, 'Most people who enjoyed Horror films also enjoyed Action movies' – 51 out of 66 of them did. However, you know from your previous work on hypothesis testing that the purpose of collecting samples is to give evidence for or against statements about the population as a whole, not just to make comments about the results that appear in the sample.

A formal version of the cinema management's question is, 'Is enjoyment of Horror films independent of enjoyment of Action movies?' you can use the sample data to investigate this question at the 1% significance level.

If enjoyment of the two types of film is independent, you would expect the probability of a randomly chosen cinema-goer enjoying both types of film to be the product of the probabilities of their enjoying each type. Formally, if p is the probability that a randomly chosen cinema-goer will enjoy Horror films and q is the probability that a randomly chosen cinema-goer will enjoy Action movies, then the joint distribution, on the assumption of independence, would be as shown in this table.

Probabilities that a cinema-goer	Enjoyed Horror films	Did not enjoy Horror films	Marginal probability
Enjoyed Action movies	pq	$(1-p)q$	q
Did not enjoy Action movies	$p(1-q)$	$(1-p)(1-q)$	$1-q$
Marginal probability	p	$1-p$	1

You can use the sample data to estimate the probabilities p and q. The number of cinema-goers in the sample who enjoyed Horror films is $51 + 15 = 66$, so the proportion of cinema-goers who enjoyed Horror films is $\frac{66}{150}$. The number of cinema-goers in the sample who enjoyed Action movies is $51 + 41 = 92$, so the proportion of cinema-goers who enjoyed Action movies is $\frac{92}{150}$.

Notice how you use the marginal totals 66 and 92 which were calculated previously.

If people enjoyed Horror films and Action movies independently with the probabilities you have just estimated, then you would expect to find, for instance,

number of people enjoying both types

$$= 150 \times P(\text{a random person enjoying both types})$$
$$= 150 \times P(\text{enjoying Horror}) \times P(\text{enjoying Action})$$
$$= 150 \times \frac{66}{150} \times \frac{92}{150}$$
$$= \frac{6072}{150}$$
$$= 40.48$$

In the same way, you can calculate the number of people you would expect to correspond to each cell in the table.

Expected numbers	Enjoyed Horror	Did not enjoy Horror	Marginal totals for Action movies
Enjoyed Action movies	$150 \times \frac{66}{150} \times \frac{92}{150} = 40.48$	$150 \times \frac{84}{150} \times \frac{92}{150} = 51.52$	92
Did not enjoy Action movies	$150 \times \frac{66}{150} \times \frac{58}{150} = 25.52$	$150 \times \frac{84}{150} \times \frac{58}{150} = 32.48$	58
Marginal totals for Horror films	66	84	150

Note that it is an inevitable consequence of this calculation that these expected figures have the same marginal totals as the sample data.

You are now in a position to test the original hypotheses, which you can state formally as:

H_0: Enjoyment of the two types of film is independent.

H_1: Enjoyment of the two types of film is not independent.

We have just calculated the expected frequencies assuming the null hypothesis is true. You know the actual sample frequencies and the aim is to decide whether they are so different that the null hypothesis should be rejected.

You have already met a statistic which measures how far apart a set of observed frequencies is from the set expected under the null hypothesis – the χ^2 (chi-squared) statistic, given by the formula:

$$X^2 = \sum \frac{(\text{observed frequency} - \text{expected frequency})^2}{\text{expected frequency}} = \sum \frac{(f_o - f_e)^2}{f_e}.$$

You can use this here: the observed and expected frequencies are summarised below.

Observed frequencies	Enjoyed Horror	Did not enjoy Horror
Enjoyed Action	51	41
Did not enjoy Action	15	43

Expected frequencies	Enjoyed Horror	Did not enjoy Horror
Enjoyed Action	40.48	51.52
Did not enjoy Action	25.52	32.48

The χ^2 statistic is:

$$X^2 = \frac{(51 - 40.48)^2}{40.48} + \frac{(41 - 51.52)^2}{51.52} + \frac{(15 - 25.52)^2}{25.52} + \frac{(43 - 32.48)^2}{32.48}$$

$$= \frac{10.52^2}{40.48} + \frac{10.52^2}{51.52} + \frac{10.52^2}{25.52} + \frac{10.52^2}{32.48} = 12.626$$

Note

The four numerators in this calculation are equal. This is not by chance; it will always happen with a 2 × 2 table – and it is worth noting, as a useful calculational short cut.

Following the usual hypothesis-testing methodology, you want to know whether a value for this statistic at least as large as 12.626 is likely to occur by chance when the null hypothesis is true. For that, you need to know the distribution of the statistic under H_0. Given its name, it should not be surprising to learn that the statistic is distributed with an approximate χ^2 distribution, of the type you used when testing 'goodness of fit' in *Statistics 3*. The large expected number of people in each cell (all greater than 25) in this case means that this approximation is very good.

The number of degrees of freedom for the χ^2 distribution is determined as in the χ^2 'goodness of fit' test: starting with the number of cells which must be filled, subtract one degree of freedom for each restriction, derived from the data, which is placed on the frequencies. Here you are imposing the requirements that the total of the frequencies must be 150, and that the overall proportions of people enjoying Horror films and Action movies are $\frac{66}{150}$ and $\frac{92}{150}$, respectively.

$v = 4$ (number of cells)

$\quad - 1$ (total of frequencies is fixed by the data)

$\quad\quad - 2$ (proportions of people enjoying each type are estimated from the data)

$= 1$

The critical value for χ_1^2 at the 1% level is 6.635. The value of X^2 obtained above, 12.626, is considerably greater than this and so you reject the null hypothesis at the 1% level and accept that people's enjoyment of the two types of film is not independent or that enjoyment of the two is *associated*.

Contingency tables in general

Suppose that two random variables, each taking values which fall into one of a finite number of discrete categories, are both measured on the elements of a sample. The table which results from listing the frequencies with which each possible pair of values of the two variables arises is called a *contingency table*. If the first variable can take m distinct values and the second can take n distinct values, it is called an $m \times n$ *contingency table*.

The 4×3 contingency table below shows the type of car (saloon, sports, hatchback or estate) owned by 360 randomly-chosen people, and the age category (under 30, 30–60, over 60) into which the owners fall.

	Age of driver			Marginal totals
	under 30	30–60	over 60	for car type
Saloon	10	67	57	134
Sports car	19	14	3	36
Hatchback	32	47	34	113
Estate	7	56	14	77
Marginal totals for age category	68	184	108	360

The marginal totals are not essential in a contingency table, but it is conventional – and convenient – to add them.

The χ^2 test for independence in a contingency table

Given the data in a contingency table it is possible to test the hypotheses:

H_0: The two variables whose values are being measured are independent in the population.

H_1: The two variables whose values are being measured are not independent in the population.

EXAMPLE 1.1

Use the data in the contingency table above to test, at the 5% significance level, the hypotheses:

H_0: Car type is independent of owner's age.

H_1: Car type is not independent of owner's age.

SOLUTION

1 You need to calculate the expected frequencies in the table if the null hypothesis is true. You use the probability estimates given by the marginal totals, for instance:

Probability car type is a hatchback $\frac{113}{360}$

Probability driver's age is over 60 $\frac{108}{360}$

The expected frequencies are then calculated as below.

Expected frequencies of hatchback and owner's age over 60

$= 360 \times \mathrm{P}(\text{car is a hatchback and owner's age over 60})$

$= 360 \times \mathrm{P}(\text{car is a hatchback}) \times \mathrm{P}(\text{owner's age over 60})$

$= 360 \times \dfrac{113}{360} \times \dfrac{108}{360} = \dfrac{113 \times 108}{360}$

> Since, under the null hypothesis, the type of car and the driver's age are independent

This illustrates the general result for contingency tables:

$$\text{expected frequency in a cell} = \frac{\text{product of marginal totals for that cell}}{\text{number of observations}}$$

The table below shows all the expected frequencies, calculated in this way.

	Age of driver			Marginal totals
	under 30	30–60	over 60	for car type
Saloon	25.311	68.489	40.200	134
Sports car	6.800	18.400	10.800	36
Hatchback	21.344	57.756	33.900	113
Estate	14.544	39.356	23.100	77
Marginal totals for age category	68	184	108	360

2 At this point it is important to check that all the frequencies are large enough to make the χ^2 distribution a good approximation to the distribution of the X^2 statistic – the usual rule of thumb is to require all the expected frequencies to be greater than 5. This requirement is (just) satisfied here – though you might be cautious in your conclusions if the X^2 statistic is very near the relevant critical value. If some of the cells have small expected frequencies, you should either collect more data, or amalgamate some of the categories if it makes sense to do so. For instance, two adjacent age ranges could reasonably be combined, but two car types probably could not.

3 The value of the X^2 statistic is, as before:

$$X^2 = \sum \frac{(\text{observed frequency} - \text{expected frequency})^2}{\text{expected frequency}}$$

$$= \frac{(10 - 25.311)^2}{25.311} + \frac{(67 - 68.489)^2}{68.489} + \ldots \text{ etc.}$$

$$= 9.262 + 0.032 + \ldots \text{ etc.}$$

Unfortunately, unlike the 2×2 special case, there is no shortcut to this calculation — the numerators are *not* equal. The table below shows the value of $\frac{(\text{observed frequency} - \text{expected frequency})^2}{\text{expected frequency}}$ for each cell.

	Age of driver		
	under 30	**30–60**	**over 60**
Saloon	9.262	0.032	7.021
Sports car	21.888	1.052	5.633
Hatchback	5.319	2.003	0.000
Estate	3.913	7.039	3.585

The sum of these values is $X^2 = 66.749$.

4 The number of degrees of freedom must be calculated: the method here appears slightly different from the one used for 'goodness-of-fit' testing, and was explained in the initial 2×2 example, but it gives the same result, as you can check.

There are $3 \times 4 = 12$ cells in the contingency table, so there are twelve possibilities for differences between the observed and expected tables: however, not all these twelve differences are independent because all the marginal totals are automatically the same in observed and expected tables. There are $3 + 4 = 7$ marginal totals, but only $7 - 1 = 6$ of these are independent, because the row totals and column totals must both give the same overall total (360 in this case). This means that the twelve initial possibilities for difference are reduced to $12 - (7 - 1) = 6$ independent possibilities which means there are 6 degrees of freedom.

In general, for an $m \times n$ table, the number of degrees of freedom is:

$$v = m \times n \longleftarrow \boxed{\text{number of cells}}$$

$$- (m + n - 1) \longleftarrow \boxed{\begin{array}{c}\text{Row and column totals are fixed}\\\text{but row totals and column}\\\text{totals have the same sum.}\end{array}}$$

$$= mn - m - n + 1$$

$$= (m - 1)(n - 1)$$

5 From the χ^2 tables, the critical value at the 5% level with six degrees of freedom is 12.59. The observed χ^2 statistic is 66.749 which is greater than the critical value and so you reject the null hypothesis and accept that car type is not independent of owner's age, or that car type and owner's age are associated.

Note

It is not obvious *how* the lack of independence arises: you have not demonstrated, for example, that increasing age of the owner is associated with less sporty cars. All you can say is that the observations were very different from what you would expect from independence of the variables, and so this independence is not plausible. However, by looking at the values of $\dfrac{(\text{observed frequency} - \text{expected frequency})^2}{\text{expected frequency}}$ for each cell calculated in **3** above you can see which differences between observed and expected frequencies are significant, in the sense of making a large contribution to the X^2 statistic. In this case, the under-30 age group own fewer saloon and estate cars, more hatchbacks and many more sports cars than expected. Other cells with relatively large contributions to the X^2 statistic correspond to estate cars being owned more often than expected by 30–60 year olds, and less often than expected by older or younger drivers, and over-60s owning more saloon cars and fewer sports cars than expected.

Yates' correction

It was pointed out above that the distribution of the X^2 statistic is only approximately χ^2: in fact, since only whole number entries are possible in the contingency table, X^2 can only take a finite number of values (for a given sample size), and its distribution must be discrete. This gives a clue to a method for improving the approximation, at least in the 2×2 case, called *Yates' correction* – it is effectively a continuity correction like those used when approximating binomial and Poisson distributions as normal ones.

Instead of calculating:
$$X^2 = \sum \frac{(\text{observed frequency} - \text{expected frequency})^2}{\text{expected frequency}}$$

use the formula:
$$X^2 = \sum \frac{(|\text{observed frequency} - \text{expected frequency}| - 0.5)^2}{\text{expected frequency}}$$

Note that it is the absolute unsigned difference that is reduced by 0.5.

EXAMPLE 1.2

Apply Yates' correction to the calculation of the X^2 value in the movie analysis.

SOLUTION

$$\frac{(|51 - 40.48| - 0.5)^2}{40.48} + \frac{(|41 - 51.52| - 0.5)^2}{51.52} + \frac{(|15 - 25.52| - 0.5)^2}{25.52}$$

$$+ \frac{(|43 - 32.48| - 0.5)^2}{32.48}$$

$$= \frac{10.02^2}{40.48} + \frac{10.02^2}{51.52} + \frac{10.02^2}{25.52} + \frac{10.02^2}{32.48} = 11.454$$

This is a relatively small change, and leaves the X^2 value still highly significant.

Note

All four of the numerators in the calculation of X^2 are still equal, so that calculations for the 2×2 contingency table can be simplified as before, even with Yates' correction. The next example shows that Yates' correction can sometimes make a very considerable difference.

EXAMPLE 1.3

In an experiment to assess the effect of an ointment on eczema 33 patients are selected and 12 of them have the ointment applied each day for a week, while 21 remain untreated. The researchers wish to test whether the ointment is effective in reducing the presence of papules. At the end of the week each patient is examined to see if there are more or fewer papules present. Test the appropriate hypothesis at the 5% level.

SOLUTION

Results	Fewer papules present	More papules present	Marginal totals for treatment
Ointment applied	9	3	12
No treatment	7	14	21
Marginal totals for outcome	16	17	33

The hypotheses being tested are:

H_0: The outcome is independent of the treatment.

H_1: The outcome is not independent of the treatment.

Expected figures	Fewer papules present	More papules present	Marginal totals for treatment
Ointment applied	$33 \times \frac{12}{33} \times \frac{16}{33} = 5.8182$	$33 \times \frac{12}{33} \times \frac{17}{33} = 6.1818$	12
No treatment	$33 \times \frac{21}{33} \times \frac{16}{33} = 10.182$	$33 \times \frac{21}{33} \times \frac{17}{33} = 10.818$	21
Marginal totals for outcome	16	17	33

So the X^2 statistic, using Yates' correction is:

$$\frac{(|9 - 5.8182| - 0.5)^2}{5.8182} + \frac{(|3 - 6.1818| - 0.5)^2}{6.1818} + \frac{(|7 - 10.182| - 0.5)^2}{10.182}$$
$$+ \frac{(|14 - 10.818| - 0.5)^2}{10.818} = 3.771$$

The critical value of the χ^2 distribution at the 5% level with one degree of freedom is 3.84, and $3.771 < 3.84$ so at the 5% level you accept the null hypothesis that the ointment has no effect.

If Yates' correction were not used the X^2 value would be calculated as 5.308 and the null hypothesis would be rejected.

Note that where Yates' correction makes a large difference to the value of X^2, the conclusion of any hypothesis test should be treated with caution. If the null hypothesis is rejected using the correction, this rejection will be justified; however, the corrected X^2 value may lead to acceptance of the null hypothesis of independence when in fact the data does support an association.

Yates' correction should only be used in 2×2 contingency tables.

Causality

Example 1.3, on eczema treatments, may have appeared to be testing something different from the others considered in this chapter. There appeared to be an asymmetry between the variables involved because you would probably expect that applying the ointment was causing the number of papules to be reduced, and certainly not the other way round. There was no such asymmetry in the movie analysis. You would not expect either of the variables, 'attitude to Horror films' and 'attitude to Action movies', to cause the other: you are just asking whether they are related (though they may be related because both are caused by a third factor – bloodthirstiness, perhaps).

Example 1.1 is less clear cut. It may or may not be true that one of the factors causing anyone to choose a sports car, for instance, is their age. It is very important to understand that from the point of view of a statistician there is no difference between these examples: the χ^2 test tells you nothing about causality. If lack of independence between two variables is supported by such a test, whether one variable causes the other (and if so, which way round the causal effect works) is a matter for separate investigation, or common sense.

It would therefore be wrong, for example, to rephrase the hypotheses of the eczema example as

H_0: The treatment does not help to cure the condition.

H_1: The treatment helps to cure the condition.

FISHER'S EXACT TEST

Recall that the X^2 statistic used above is discrete, and has a distribution which is only approximated by the χ^2 distribution. This investigation is about the exact distribution of X^2.

The table below shows a set of observed data, where the variables A, taking values a and a', and B, taking values b and b', have been measured on a sample of size 25. The hypothesis to be tested is that A and B are independent.

	$A = a$	$A = a'$	Marginal totals for B
$B = b$	2	10	12
$B = b'$	9	4	13
Marginal totals for A	11	14	25

Investigate the critical region for this test: for the same marginal totals, which values in the body of the table are unlikely to occur by chance if A and B are independent?

1 Verify that, with the given marginal totals, the entry in the $A = a$, $B = b$ cell can take values $0, 1, \ldots, 11$, and that once the entry in this cell is chosen, the entries in the other three cells are determined. Show that this can be expressed algebraically as:

	$A = a$	$A = a'$	Marginal totals for B
$B = b$	r	$12 - r$	12
$B = b'$	$11 - r$	$2 + r$	13
Marginal totals for A	11	14	25

where r can take values $0, 1, \ldots, 11$.

2 Calculate the X^2 statistic for this table in terms of r and show that your result simplifies to:

$$\frac{15\,625}{24\,024}\left(r - \frac{132}{25}\right)^2$$

3 Hence, explain why, if you test at the 5% level, the critical region, assuming an approximate χ^2 distribution for X^2 is:

$$\frac{15\,625}{24\,024}\left(r - \frac{132}{25}\right)^2 > 3.841$$

and show that this simplifies to a critical region of:

$$\{r < 2.85\} \cup \{r > 7.71\}$$

or:

$$\{0, 1, 2\} \cup \{8, 9, 10, 11\}$$

4 The exact probabilities for each possible value of r that you must calculate are:
P(the $A = a$, $B = b$ cell contains r elements | the marginal totals are correct)
and you can do this combinatorially, assuming that the null hypothesis holds, by
imagining that the 25 elements of the sample are being randomly distributed
amongst the four cells. Explain why the probability you want is:

$$\frac{n(\text{the } A = a, \ B = b \text{ cell contains } r \text{ elements and the marginal totals are correct})}{n(\text{the marginal totals are correct})}$$

where the notation $n(\ldots)$ means the number of possible arrangements of the
elements of the sample where $\ldots$ holds.

5 Explain why:
$n(\text{the } A = a, B = b \text{ cell contains } r \text{ elements and the marginal totals are correct})$

$$= \binom{25}{12} \times \binom{12}{r} \times \binom{13}{11 - r}$$

and

$$n(\text{the marginal totals are correct})$$

$$= \binom{25}{12} \times \binom{25}{11}$$

so that
P(the $A = a$, $B = b$ cell contains r elements | the marginal totals are correct)

$$= \frac{\binom{12}{r} \times \binom{13}{11 - r}}{\binom{25}{11}}$$

6 Evaluate these probabilities to complete the table started below.

r	P(the $A = a, B = b$ cell contains r elements | the marginal totals are correct)
0	0.000 017 5
1	0.000 770 0
$\ldots$	
10	0.000 192 5
11	0.000 002 7

Hence show that:

$$P(r \in \{0, 1, 2\} \cup \{8, 9, 10, 11\}) = 0.047\ 18 < 0.05$$

but that no larger set of extreme r-values has probability less than 5%, so that the
χ^2 approximation gives a correct critical region for the test in this case.

7 Repeat this calculation for smaller sample sizes, or less evenly-balanced marginal
totals, where the expected frequencies are not all greater than 5. When does the
approximation break down?

1 A medical insurance company office is the largest employer in a small town. When 37 randomly-chosen people living in the town were asked where they worked and whether they belonged to the town's health club, 21 were found to work for the insurance company, of whom 15 also belonged to the health club, while 7 of the 16 not working for the insurance company belonged to the health club.

Test the hypothesis that health club membership is independent of employment by the medical insurance company.

2 A group of 281 voters is asked to rate how good a job they think the Prime Minister is doing. Each is also asked for the highest educational qualifications they have achieved. The frequencies with which responses occurred are shown in the table.

Highest qualifications achieved

Rating of PM	None	GCSE or equivalent	A-level or equivalent	Degree or equivalent
Very poor	11	37	13	6
Poor	12	17	22	8
Moderate	7	11	25	10
Good	10	17	17	9
Very good	19	16	8	6

Use these figures to test whether there is an association between rating of the Prime Minister and highest educational qualifications achieved.

3 In a random sample of 163 adult males, 37 suffer from hay-fever and 51 from asthma, both figures including 14 men who suffer from both. Test whether the two conditions are associated.

4 In a survey of 184 London residents brought up outside the south east of England, respondents were asked whether, job and family permitting, they would like to return to their area of origin. Their responses are shown in the table.

Region of origin	Would like to return	Would not like to return
South-west	16	28
Midlands	22	35
North	15	31
Wales	8	6
Scotland	14	9

Test the hypothesis that desire to return is independent of region of origin.

5 A sample of 80 men and 150 women selected at random are tested for colour-blindness. Twelve of the men and five of the women are found to be colour-blind. Is there evidence at the 1% level that colour-blindness is sex-related?

6 Depressive illness is categorised as type I, II or III. In a group of depressive psychiatric patients, the length of time for which their symptoms are apparent is observed. The results are shown below.

Length of depressive episode	Type of symptoms		
	I	II	III
Brief	15	22	12
Average	30	19	26
Extended	7	13	21
Semi-permanent	6	9	11

Is the length of the depressive episode independent of the type of symptoms?

7 The personnel manager of a large firm is investigating whether there is any association between the length of service of the employees and the type of training they receive from the firm. A random sample of 200 employee records is taken from the last few years and is classified according to these criteria. Length of service is classified as short (meaning less than 1 year), medium (1–3 years) and long (more than 3 years). Type of training is classified as being merely an initial 'induction course', proper initial on-the-job training but little if any more, and regular and continuous training. The data are as follows:

Type of training	Length of service		
	Short	Medium	Long
Induction course	14	23	13
Initial on-the-job	12	7	13
Continuous	28	32	58

Examine at the 5% level of significance whether these data provide evidence of association between length of service and type of training, stating clearly your null and alternative hypotheses.

Discuss your conclusions.

8 The bank manager at a large branch was investigating the incidence of bad debts. Many loans had been made during the past year; the manager inspected the records of a random sample of 100 loans, and broadly classified them as satisfactory or unsatisfactory loans and as having been made to private individuals, small businesses or large businesses. The data were as follows.

	Satisfactory	Unsatisfactory
Private individual	22	4
Small business	34	13
Large business	24	4

Carry out a χ^2 test at the 5% level of significance to examine if there is any association between whether or not the loan was satisfactory and the type of customer to whom the loan was made. State clearly the null and alternative hypotheses and the critical value of the test statistic.

9 In the initial stages of a market research exercise to investigate whether a proposed advertising campaign would be worthwhile, a survey of newspaper readership was undertaken. 100 people selected at random from the target population were interviewed. They were asked how many newspapers they read regularly. They were also classified as to whether they lived in urban or rural areas. The results were as follows:

Number of newspapers read regularly	Urban	Rural
None	15	11
One	22	18
More than one	27	7

(i) Examine at the 10% level of significance whether these data provide evidence of an association between the categories. State clearly the null and alternative hypotheses you are testing.

(ii) Justify the number of degrees of freedom for the test.

(iii) Discuss the conclusions reached from the test.

10 Public health officers are monitoring air quality over a large area. Air quality measurements using mobile instruments are made frequently by officers touring the area. The air quality is classified as poor, reasonable, good or excellent. The measurement sites are classified as being in residential areas, industrial areas, commercial areas or rural areas. The table shows a sample of frequencies over an extended period. The row and column totals and the grand total are also shown.

Measurement site	Air quality				
	Poor	Reasonable	Good	Excellent	Row totals
Residential	107	177	94	22	400
Industrial	87	128	74	19	308
Commercial	133	228	148	51	560
Rural	21	71	24	16	132
Column totals	348	604	340	108	1400

Examine at the 5% level of significance whether or not there is any association between measurement site and air quality, stating carefully the null and alternative hypotheses you are testing. Report briefly on your conclusions.

11 (i) A large co-educational school provides a cafeteria lunch service and also allows students to bring their own packed lunches. A survey was conducted to investigate whether or not there was any connection between a student's sex and his or her use of the cafeteria. A random sample of 50 students was chosen; 26 were boys of whom 15 used the cafeteria and the remainder brought packed lunches; of the 24 girls, there were only 7 who used the cafeteria, the remainder bringing packed lunches. Display the data in a 2×2 contingency table and calculate the expected frequencies on the hypothesis that there is no such connection.

(ii) Carry out the usual χ^2 test, at the 5% level of significance, *without using Yates' correction* in your calculations, and interpret the result.

(iii) Repeat (ii) but this time *using Yates' correction* in the calculations.

(iv) In what circumstances could use of Yates' correction affect the conclusion of a χ^2 test?

(v) Your results from (ii) and (iii) should have led you to different conclusions. How would you attempt to reconcile these in presenting the results of the survey?

[MEI]

12 In an investigation of small business development in England, researchers are examining whether there is any association between the geographical area where such a business is located and the life-span of the business. A random sample of records has been obtained from a national database. The geographical areas are classified very broadly as South-east, Midlands, North and 'Rest'. Lifespans are classified as short, medium and long. The table shows the frequencies obtained in the sample; row and column totals and the grand total are also shown.

Geographical area	Lifespan of business			Row totals
	Short	Medium	Long	
South-east	140	72	56	268
Midlands	104	53	45	202
North	71	51	48	170
Rest	57	48	59	164
Column totals	372	224	208	804

Examine at the 1% level of significance whether or not there is any association between geographical area and lifespan, stating carefully the null and alternative hypotheses you are testing. Report briefly on your conclusions.

[MEI]

Testing whether data come from a given normal distribution

It is a common assumption in statistics that a set of data arises from a normal distribution, for instance when conducting a t-test. How can you test, from the information in a sample, whether this assumption is justified?

When an intelligence test was standardised, scores on the test were distributed normally with mean 100 and standard deviation 15. Twenty years later, it is thought that the distribution of scores may have changed. If Q is the random variable giving an individual's test score, test the hypotheses:

H_0: Q has distribution $N(100, 15^2)$.

H_1: Q does not have the distribution $N(100, 15^2)$.

The intelligence scores of a random sample of 40 people are given below.

92	106	91	112	106	113	125	108
103	127	110	112	120	97	115	90
119	87	114	90	88	117	119	108
103	94	104	116	97	112	103	97
86	82	114	120	115	94	110	106

One possible way of grouping these data is as follows.

Intelligence	Frequency
–87.5	3
87.5–96.5	7
96.5–103.5	6
103.5–112.5	11
112.5–	13

You can now calculate the frequencies you would expect in these intervals under the assumption that the null hypothesis is true.

You know that Q is a random variable for which the distribution is $N(100, 15^2)$ under the null hypothesis, so that:

$$P(87.5 < Q < 96.5) = P\left(\frac{87.5 - 100}{15} < Z < \frac{96.5 - 100}{15}\right)$$

$$= P(-0.8333 < Z < -0.2333)$$

$$= \Phi(0.8333) - \Phi(0.2333) = 0.7976 - 0.5923 = 0.2053$$

and so the expected frequency for the interval 87.5–96.5 is $0.2053 \times 40 = 8.212$.

The other expected frequencies can be found by similar calculations (and the symmetry of the intervals about the hypothesised mean), resulting in the table of observed and expected frequencies below.

Range of IQ scores	Observed frequency	Expected frequency
–87.5	3	8.096
87.5–96.5	7	8.212
96.5–103.5	6	7.384
103.5–112.5	11	8.212
112.5–	13	8.096

It should now be no surprise that the X^2 statistic

$$\sum \frac{(\text{observed frequency} - \text{expected frequency})^2}{\text{expected frequency}}$$

has an approximately χ^2 distribution under the null hypothesis. Its number of degrees of freedom is calculated as for the 'goodness-of-fit' test for discrete distributions that was discussed in *Statistics 3*:

$$v = \text{number of classes} - 1$$

where one degree of freedom is lost because the totals of expected and observed frequencies are equal.

Here:

$$X^2 = \frac{(3 - 8.096)^2}{8.096} + \frac{(7 - 8.212)^2}{8.212} + \frac{(6 - 7.384)^2}{7.384} + \frac{(11 - 8.212)^2}{8.212}$$
$$+ \frac{(13 - 8.096)^2}{8.096}$$
$$= 7.563$$

and $v = 5 - 1 = 4$

Observe that the expected frequencies for each cell are greater than 5, as required for the χ^2 distribution to be a good approximation, so that you can use the χ^2 tables to determine the critical value: for four degrees of freedom at the 5% level this is 9.488. Since $7.563 < 9.488$, you can accept the null hypothesis that the sample is drawn from an underlying $N(100, 15^2)$ distribution.

Note

It is very important to be clear exactly what the acceptance of the null hypothesis means: that it is not particularly implausible that the data seen could have arisen from random sampling of the stated normal distribution. In no sense have you confirmed that the underlying distribution does have this form, merely that it is not unreasonable to assume that it does. For instance, the same data used in a *t*-test to test the null hypothesis $\mu = 100$ on the *assumption* that the sample is drawn from a normal distribution with mean μ leads to rejection of the null hypothesis at the 1% level.

Testing for normality without a known mean and variance

When testing for normality of the underlying distribution, in preparation for conducting a t-test for instance, you are merely asking whether it is appropriate to assume that the underlying distribution is normal in shape; not whether it has a specific mean and variance.

An experiment is conducted to determine whether people's estimates of one minute have a mean duration of one minute. Data is to be collected by asking a sample of people to say 'Start' and 'Stop' at times they estimate to be one minute apart. The actual time apart, in seconds, is recorded by the experimenter. A t-test is to be conducted of the hypothesis that the mean actual time apart is 60 seconds. Before this is done, a preliminary sample is taken to test the hypotheses:

H_0: Estimates are normally distributed.

H_1: Estimates are not normally distributed.

The estimates (in seconds) obtained when this preliminary sample was taken are listed below.

55	40	50	53	57	61	38	29	43	52
37	57	55	56	57	48	59	40	54	53
63	58	56	48	55	58	57	56	59	55
50	60	58	51	42	47	62	57	49	43
51	42	39	56	53	53	58	51	50	55
40	38	41	55	45	61	53	53	41	53

From these data, calculate the sample mean and the usual sample estimate of the population standard deviation:

$$\bar{x} = 51.1 \text{ and } s = \sqrt{\frac{\sum(x - \bar{x})^2}{n - 1}} = 7.564$$

Use these estimated parameters to calculate the expected frequencies; that is, test the fit of the data to the normal distribution $N(51.1, 7.564^2)$.

The data must now be grouped. One possible grouping is shown below.

Estimates (seconds)	Frequency
−41.5	10
41.5–45.5	5
45.5–49.5	4
49.5–53.5	14
53.5–57.5	16
57.5–	11

The expected frequencies for these groups can be calculated as set out below.

Class	Upper class boundary	Standardised value ζ	$P(Z < \zeta)$	Probability for class	Expected frequency
−41.5	41.5	−1.2692	0.1022	0.1022	6.132
41.5–45.5	45.5	−0.7404	0.2295	0.1273	7.638
45.5–49.5	49.5	−0.2115	0.4162	0.1867	11.202
49.5–53.5	53.5	0.31729	0.6244	0.2082	12.492
53.5–57.5	57.5	0.84611	0.8012	0.1768	10.608
57.5–	∞	∞	1	0.1988	11.928

The X^2 statistic is therefore:

$$\frac{(10 - 6.132)^2}{6.132} + \frac{(5 - 7.638)^2}{7.638} + \frac{(4 - 11.202)^2}{11.202} + \frac{(14 - 12.492)^2}{12.492}$$
$$+ \frac{(16 - 10.608)^2}{10.608} + \frac{(11 - 11.928)^2}{11.928}$$
$$= 2.440 + 0.911 + 4.630 + 0.182 + 2.741 + 0.072 = 10.976$$

To calculate the number of degrees of freedom, recall that you have used the data to estimate two parameters (mean and standard deviation) for the distribution. This means that both the observed and expected frequencies must give the same total frequency, the same sample mean and the same sample estimate of the standard deviation. These three restrictions on the possible frequencies in each class reduce the number of degrees of freedom by three from the number of classes.

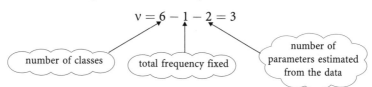

$$\nu = 6 - 1 - 2 = 3$$

number of classes — total frequency fixed — number of parameters estimated from the data

With three degrees of freedom, the critical value at the 5% level for the χ^2 distribution is 7.815.

Since $12.23 > 7.815$, you reject the null hypothesis that the data were drawn from a normally distributed population, and it would not be appropriate to use a t-test for assessing the hypothesis $\mu = 60$, although with a large enough sample an appeal to the central limit theorem would allow you to use a normal test.

As with the contingency table earlier, from the values of $\dfrac{(\text{observed} - \text{expected})^2}{\text{expected}}$ for each cell calculated above you can see which differences between observed and expected frequencies are significant, in the sense of making a large contribution to the X^2 statistic. The observed frequency of 10 in the first cell is itself substantially higher than the expected frequency of 6.132, but also this excess of low estimates also brings down our estimate of the mean to the lower end of the main peak of the distribution, hence the class above that containing the estimated mean has a significantly higher observed than expected frequency and the class below that containing the estimated mean a significantly lower observed than expected frequency.

Note

1 When testing for an underlying normal distribution as a prelude to conducting a *t*-test it is important that different samples are used for the preliminary test of the distributional assumption and the actual test: this is because the sampling distribution of the *t*-statistic in the two cases:

    (i) the population is normally distributed and a sample is taken at random

    (ii) the population is normally distributed and a sample is taken at random except for the restriction that it 'passes' the χ^2 test for normality,

are not the same.

2 There is a certain amount of arbitrariness in the grouping of data which precedes the 'goodness-of-fit' test: you want to ensure that there are enough classes to discriminate between different distributions, but that each is wide enough to have an expected frequency of at least five. There is no need to choose constant class widths, and in fact it would be wise to have narrower classes where the expected distribution has the greatest density. Picking class widths so that the expected frequencies are all about 8–12 is a reasonable rule of thumb.

Testing goodness of fit with other continuous distributions

The goodness of fit of a set of data to a continuous underlying distribution other than the normal can also be tested using the χ^2 test technique, as the discussion below illustrates.

The time intervals between arrivals of passengers at a bus stop might be modelled by an exponential distribution, provided that buses appear sufficiently frequently and unpredictably for passengers simply to turn up independently at random.

Test the hypotheses:

H_0: The time intervals between passenger arrivals are exponentially distributed.

H_1: The time intervals between passenger arrivals are not exponentially distributed.

The time intervals between 36 successive arrivals at a bus stop were measured (in seconds) and recorded in the table below.

60	43	25	25	31	37	23	23	14
13	36	3	63	50	33	18	60	52
6	38	41	33	52	28	36	30	42
16	10	55	161	21	1	3	14	13

The mean of these data is 33.58.

A grouped frequency distribution for the sample looks like this.

Time interval (nearest second)	0–9	10–19	20–29	30–39	40–49	50–59	60–69	70–79
Frequency	4	7	6	8	3	4	3	1

To calculate the expected frequencies, you need to use the density function for the exponential distribution. This is given by:

$$f(t) = \lambda e^{-\lambda t}$$

The mean of this distribution is $\dfrac{1}{\lambda}$ so you can estimate λ from the sample mean as

$$\frac{1}{\bar{x}} = 0.029\,78.$$

The probability that an exponentially distributed variable lies between a and b is:

$$\int_a^b \lambda e^{-\lambda t} dt = \left[-e^{-\lambda t}\right]_a^b = e^{-\lambda a} - e^{-\lambda b}$$

so the expected frequencies can be calculated, for example, for the class interval 20–29 (i.e. with class boundaries 19.5 and 29.5) as:

$$36 \times \left(e^{-0.02978 \times 19.5} - e^{-0.02978 \times 29.5}\right) = 36 \times 0.144\,10 = 5.1874$$

The complete set of expected frequencies is:

Time interval (nearest second)	Probability	Expected frequency
0–9	0.246 39	8.8700
10–19	0.194 07	6.9867
20–29	0.144 10	5.1874
30–39	0.106 99	3.8515
40–49	0.079 44	2.8597
50–59	0.058 98	2.1232
60–69	0.043 79	1.5764
70–	0.126 25	4.5451

Note that the expected frequency is calculated for the entire interval 70–∞, not just the interval 70–79 in which the maximum of the actual data lies. In conducting a χ^2 test, it is essential that the intervals for which the expected frequency are calculated cover the whole range of the theoretical distribution – not just the range of the actual data. An alternative way of doing this would have been to add an extra interval 80–∞, with observed frequency 0. However, in this example, the expected frequency would have been considerably less than 5, so the figures used to calculate the final X^2 statistic would have been the same. These expected frequencies are not all greater than 5, so the class boundaries need to be redrawn. One way of doing this is as follows (overleaf).

Time interval (nearest second)	Observed frequency	Expected frequency
0–9	4	8.8700
10–19	7	6.9867
20–29	6	5.1874
30–49	11	6.7112
50–	8	8.2447

The X^2 statistic is

$$X^2 = \frac{(4 - 8.8700)^2}{8.8700} + \frac{(7 - 6.9867)^2}{6.9867} + \frac{(6 - 5.1874)^2}{5.1874} + \frac{(11 - 6.7112)^2}{6.7112}$$
$$+ \frac{(8 - 8.2447)^2}{8.2447}$$
$$= 5.549$$

and the number of degrees of freedom is:

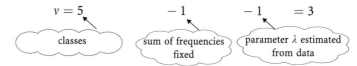

$$\nu = 5 \qquad -1 \qquad -1 \qquad = 3$$

classes   sum of frequencies fixed   parameter λ estimated from data

The critical value at the 5% level for three degrees of freedom is 7.815.

Since $5.549 < 7.815$ you accept the null hypothesis that the underlying distribution from which the data are drawn is exponential.

INVESTIGATION

Generate data from a rectangular distribution on [0, 1], using a calculator or computer. With sample size n, test whether the data you have generated could be taken as arising from the normal distribution

$$N\left(\frac{1}{2}, \frac{1}{12}\right)$$

(which has the same mean and variance as the rectangular distribution on [0, 1]).

How large does n have to be before the null hypothesis is reliably rejected?

What does this tell you about how powerful the χ^2 test is?

Repeat for other pairs of distributions.

1 The time intervals (in minutes) between 25 successive buses arriving at the stop whose passenger arrivals were investigated in the text were recorded as shown below.

10.7	0.4	6.8	1.4	2.0
8.8	3.1	1.8	1.0	5.5
11.5	8.4	1.0	1.1	5.7
0.4	11.7	5.0	2.8	2.2
4.5	3.1	4.4	6.3	5.5

(i) Test whether these data can be taken as arising from an exponential distribution.

(ii) Why would you expect an exponential model to be less appropriate for the inter-arrival times of buses than passengers?

2 Fifty-five people are asked to estimate a one-metre length, by marking off their estimate on a blank straight edge. The actual length marked off is then recorded.

(i) The first 45 results are given below, in centimetres.

112	109	89	110	116	99
109	120	132	80	95	101
107	142	110	111	76	89
100	103	132	117	121	112
110	126	105	98	108	80
87	97	116	126	104	110
128	103	88	118	72	77
87	117	126			

(a) Test whether the estimates can be taken as arising from a normal distribution.

The remaining ten results, given below in centimetres, are to be used to test whether the mean estimate is 1 metre, using a *t*-test.

114	115	118	120	98	117
81	91	107	115		

(b) State whether such a test is appropriate, and if so, carry out the test.

(ii) Use all 55 results to carry out a χ^2 test for goodness of fit to the distribution $N(100, \sigma^2)$, where σ^2 is to be estimated from the data.

(iii) Comment on the difference between the procedures in **(i)** and **(ii)**. What has been tested in each case?

3 It is thought that the lifetime of a miniaturised pump component has a distribution given by the density function:

$$f(t) = \frac{\pi}{2\mu^2} t e^{-\left(\frac{\pi t^2}{4\mu^2}\right)} \ (0 \leqslant t < \infty)$$

where μ is the mean lifetime of the component.

(i) Use the following data, which give lifetimes (in hours) of 30 randomly-selected components, to test the hypothesis that the lifetimes might indeed have this distribution, estimating the value of μ by the sample mean.

181	135	142	184	102	82
80	187	160	149	63	77
118	130	74	75	100	140
163	214	146	112	92	179
150	150	107	112	169	133

(ii) Test instead whether the data could be taken as arising from a normal distribution.

4 A market researcher is supposed to stop every second passer-by and ask a question. The researcher records the time at which each person is stopped as well as the response to the question. The researcher's supervisor suspects that the researcher is not carrying out instructions and decides to test whether the times between questions fit the appropriate distribution, which, if passers-by arrive independently and at random at a constant uniform rate, is given by a gamma distribution, with density function:

$$f(t) = \lambda^2 t e^{-\lambda t}$$

with parameter λ.

The supervisor's data for the times between 80 successive questions are listed below.

7.8	10.4	3.9	4.7	5.5	2.3
13.6	8.2	21.4	5.3	15.8	5.7
18.1	7.8	11.5	8.5	3.3	6.3
9.5	9.9	9.3	5.3	2.3	9.0
10.1	14.0	13.0	8.0	8.9	13.5
6.7	3.3	9.3	13.2	7.3	1.9
0.8	10.4	4.7	15.2	3.4	16.5
6.7	6.9	5.0	5.7	22.2	7.0
4.9	8.4	4.4	7.2	8.2	4.2
12.5	16.0	8.3	9.1	3.3	1.1
7.3	9.3	5.7	10.0	14.5	4.7
2.8	2.9	6.3	9.1	3.7	23.4
10.3	9.0	10.5	5.5	4.8	10.0
15.2	7.1				

(i) Show that the mean of the gamma distribution is $\dfrac{2}{\lambda}$ and hence use the sample mean to estimate an appropriate value for λ.

(ii) Use this value of λ to test whether the data are fitted by the gamma distribution.

(iii) If the researcher was stopping every third passer-by, the appropriate gamma distribution would have density function:

$$f(t) = \frac{\lambda^3 t^2}{2} e^{-\lambda t}$$

$\left(\text{The mean of this distribution is } \dfrac{3}{\lambda}.\right)$

Are the data fitted by this distribution?

(iv) Comment.

5 The adult Heath Blue butterflies of Northern France have wing-tip to wing-tip widths which are normally distributed with mean width 31.2 mm and standard deviation 2.8 mm. The butterfly is extinct in Britain and, as part of a plan to re-introduce it from France, a lepidopterist wants to know whether the British population of the past is identical with the current French population. She has a sample of adult British butterflies from a museum, and proposes to test whether their wing-tip to wing-tip widths are well modelled by the same normal distribution as the current French population.

The wing-tip to wing-tip widths of the museum sample are given in the frequency table below.

Width class (mm)	Frequency
25–	1
26–	2
27–	0
28–	0
29–	4
30–	1
31–	8
32–	4
33–	6
34–	3
35–	1
36–	4
37–	2
38–	2
39–40	1

Carry out the lepidopterist's test.

6 The daily number of hours of sunshine at a weather station during spring is thought to be modelled by the density function:

$$f(t) = \begin{cases} k(12 - t)(4t^2 - 33t + 90) & 0 \leqslant t \leqslant 12 \\ 0 & \text{otherwise} \end{cases}$$

(i) Sketch the density function and explain the relevance of its shape to the variable being modelled.

(ii) Show that k must have the value $\dfrac{1}{3888}$.

(iii) The data below show the number of hours of sunshine recorded on randomly chosen spring days at the station. Use these data to test whether the model is appropriate.

t-values (hours)	Frequency
$0 \leqslant t < 1$	5
$1 \leqslant t < 2$	10
$2 \leqslant t < 3$	4
$3 \leqslant t < 4$	3
$4 \leqslant t < 5$	2
$5 \leqslant t < 6$	1
$6 \leqslant t < 7$	0
$7 \leqslant t < 8$	4
$8 \leqslant t < 9$	8
$9 \leqslant t < 10$	2
$10 \leqslant t < 11$	4
$11 \leqslant t < 12$	2

(iv) Explain why, if the data were recorded on 45 consecutive days during one spring, one of the assumptions of your hypothesis test would be invalid.

7 The electrical resistances of 120 inductor coils are measured, with results (in ohms) as shown below.

Range of resistances (ohms)	Frequency
160–	2
170–	5
180–	6
190–	3
200–	0
210–	11
220–	19
230–	27
240–	15
250–	17
260–	8
270–	3
280–	0
290–	3
300–310	1

Test the hypothesis that the population of inductors, from which this sample was drawn, have resistances which are normally distributed.

8 A city bus route runs through very congested streets and it is not practical to expect buses to keep to a timetable. The bus manager is endeavouring to investigate the running of these buses by modelling the times X (in minutes) between successive arrivals at a monitoring point as a continuous random variable having probability density function

$$f(x) = \begin{cases} \lambda e^{-\lambda x} & x > 0 \\ 0 & \text{elsewhere.} \end{cases}$$

where $\lambda > 0$.

(i) Obtain the mean of X.

(ii) Deduce that a plausible estimate of λ is $(\bar{x})^{-1}$ where $\bar{x}$ denotes the mean of a random sample of observations from X.

Data are available showing the intervals between one hundred successive arrivals at the monitoring point. These data, assumed to be a random sample, are summarised as follows:

Time between successive arrivals (minutes)	0–3	3–10	>10
Frequency in this category	46	42	12
Average time (minutes) between arrivals in this category	1	5	12

(iii) Show that the estimate of λ obtained as in part **(ii)** is 0.25.

(iv) Use a χ^2 test to examine the fit of the model to the data, using a 5% significance level.

9 The continuous random variable X has probability density function

$$f(x) = \begin{cases} \alpha(\alpha + 1)x^{\alpha-1}(1 - x) & 0 \leqslant x \leqslant 1 \\ 0 & \text{elsewhere} \end{cases}$$

where α is a parameter.

(i) Show that $\mu = \dfrac{\alpha}{\alpha + 2}$ where μ is the mean of X.

(ii) Deduce that a reasonable estimate of α is $a = \dfrac{2\bar{x}}{1 - \bar{x}}$ where $\bar{x}$ is the mean of a random sample of observations from X.

(iii) A random sample of 100 observations is to be tested to see whether the random variable X is a reasonable model for the underlying population. The data are given as the following grouped frequency distribution:

Range	$0 \leqslant x < 0.2$	$0.2 \leqslant x < 0.6$	$0.6 \leqslant x \leqslant 1.0$
Frequency	16	56	28

Use the value of $\bar{x}$ as given by this frequency distribution to obtain the value 1.73 for a.

(iv) Using this value of a show that the estimated expected frequency in the range $0 \leqslant x < 0.2$ for 100 random observations from X is 14.7.

(v) Given that the corresponding frequencies for $0.2 \leqslant x < 0.6$ and $0.6 \leqslant x \leqslant 1.0$ are 55.2 and 30.1, respectively, use a χ^2 test to examine at the 5% level of significance the goodness of fit of X to the data.

10 A random sample of 60 independent observations is under investigation. It is desired to examine whether a reasonable model for the underlying population is the continuous random variable X whose probability density function is

$$f(x) = \lambda e^{-\lambda x} \quad x \geqslant 0$$

where λ is a parameter $(\lambda > 0)$.

(i) Show that
$$P(a \leqslant X < b) = e^{-\lambda a} - e^{-\lambda b} \text{ where } b > a > 0.$$

(ii) Using the fact that $E(X) = \dfrac{1}{\lambda}$, explain why a reasonable estimate of λ is $\dfrac{1}{\bar{x}}$ where $\bar{x}$ is the mean of a random sample of observations of X.

(iii) The random sample of 60 observations under investigation is recorded as the following frequency distribution:

Range	$0 \leqslant x < 5$	$5 \leqslant x < 10$	$10 \leqslant x < 15$	$15 \leqslant x < 20$
Frequency	14	16	16	14

Use the value of $\bar{x}$ given by this frequency distribution to obtain the value $\frac{1}{10}$ for the estimate of λ.

(iv) Using $\frac{1}{10}$ as an estimate of λ, set up the appropriate table of estimated expected frequencies.

(v) Use a χ^2 test at the 5% level of significance to show that the model X appears not to fit the data well.

(vi) Discuss briefly the principal differences between the model X and the data.

1 An $m \times n$ *contingency table* results when two variables are measured on a sample with the first variable having m possible categories of results and the second variable having n possible categories of results. Each *cell* of the table contains an *observed frequency* with which that pair of categories of values of the two variables occurs in the sample.

2 To test whether the variables in an $m \times n$ contingency table are independent the steps are as follows.

- The null hypothesis is that the variables are independent, the alternative is that they are not.

- Calculate the marginal (row and column) totals for the table.

- Calculate the *expected frequency* in each cell as

$$\frac{\text{row total} \times \text{column total}}{\text{total sample size}}$$

- The X^2 statistic is

$$\sum \frac{(f_o - f_e)^2}{f_e}$$

where f_o is the observed frequency and f_e the expected frequency in each cell.

- The number of *degrees of freedom* for the test is $(m-1)(n-1)$ for an $m \times n$ table.

- Read the critical value from the χ^2 tables for the appropriate number of degrees of freedom and significance level. If X^2 is less than the significance level, the null hypothesis is accepted; otherwise it is rejected.

- If two variables are not independent you say that there is an *association* between them.

3 To test whether a set of data is well modelled by a random variable with a continuous distribution the steps are as follows:

- The null hypothesis is that the random variable models the data well, the alternative is that it does not.

- Group the data into *classes* and record the observed frequency in each class.

- If necessary, use the observed data to estimate one or more parameters of the hypothesised distribution.

- Add or extend classes as necessary so that the classes cover the whole range of the theoretical distribution.

- Calculate the *expected frequency* in each class, using the hypothesised distribution to find the expected probabilities, and then multiplying by the total observed frequency.

- Check whether the expected frequency in each class is greater than about 5. If not, amalgamate adjacent classes in the tails until this is true.

- The X^2 statistic is

$$\sum \frac{(f_o - f_e)^2}{f_e}$$

where f_o is the observed frequency and f_e the expected frequency in each class.

- The number of *degrees of freedom* for the test is:

(the number of classes) − (the number of estimated parameters) − 1

- Read the critical value from the χ^2 tables for the appropriate number of degrees of freedom and significance level. If X^2 is less than the significance level, the null hypothesis is accepted; otherwise it is rejected.

2 Modelling

To understand God's purpose we must study statistics for these are the measure of his purpose.

Florence Nightingale

Does the gender of a criminal affect the severity of the sentence imposed?

How might a statistician investigate this question?

The first stage would be to model the situation mathematically. You might suggest that for each convicted offender there is a constant probability of probation, suspended sentence or imprisonment, but that these probabilities differ for men and women.

Secondly, you would want to collect some data. You would need to decide how to select a sample of convicted men and women and the sentences imposed on them.

Then, you could use the data to investigate the question asked. You could estimate the probabilities in the suggested model by the relative frequencies with which the three types of sentence are imposed on men and women. A χ^2 test might then be appropriate, testing the hypothesis that the severity of the sentence is independent of the sex of the offender.

Finally, as with all mathematical modelling, you would check whether the model is adequate for your purposes or whether it needs revising. Does the sampling method justify the assumptions made in the χ^2 test? Is there a confounding variable (such as the fact that men are more likely to have previous convictions) which will produce a spurious association? Is the sample size adequate to average out the other effects on severity of sentence?

Statisticians are interested in investigating *data generating processes*, the actual mechanisms by which outcomes arise in the real world. For example:

- all the information brought out at the trial, the prejudices of the judge and the current climate of public opinion which influence the severity of the sentence

- the operator inattention, machine wear or inadequate servicing which lead to faulty devices appearing from a production line

- the combination of genetic endowment, maternal fitness and infant nurture which govern the state of development of a four-month-old puppy

- the angle and height of the roll, the nature of the table surface and the air currents in the room, which determine the position in which a die lands.

Notice that no mention has been made of randomness, and this is deliberate. Actual events do not occur by chance: they occur as a consequence of the state the world is in, and of the actions and decisions taken in this state. They are unpredictable because our knowledge of the world is insufficiently complete to enable us to decide what exactly is going to happen and it is impractical, or too costly, to find the necessary information.

Note

The physical theory of quantum mechanics suggests that in some very small-scale aspects of the universe, such as radioactive beta decay, chance is genuinely involved. This will not be relevant in any of the situations studied in this Statistics course.

The job of the statistician, like that of any applied mathematician, is to provide a useful model of the real world. All mathematical models are simplified representations of reality, ignoring complexities which will have unimportant effects on the final outcome. The unique feature of statistical modelling is that it uses randomness to model those parts of a situation where the details of the process which generate the outcome are unknown, assigning probabilities to possible outcomes, rather than predicting definitely which will occur.

Modelling a situation statistically rather than practically or by calculation is a choice. It is not correct to say that a situation is modelled statistically 'when it involves random chance': the decision to assign probabilities to outcomes is part of the modelling process. For instance, if you wanted to model the position at which a dart struck a dartboard, you could consider the mechanical equations governing the motion of a complex projectile, perhaps ignoring air resistance, and arrive at an equation linking the striking point to the initial speed and angle of projection. Alternatively, you could suggest the probabilities with which the dart would land in the various sectors of the board, given the sector at which it was aimed. Which of these would be a better model would depend on the reason for your interest in the outcomes, on how much trouble you were prepared to go to for a given level of accuracy in prediction, and on which factors most affect the striking point of the dart. It does not depend on whether or not there 'really is' a random aspect to the motion of the dart.

The construction of a statistical model involves:

A Focusing attention on a particular aspect of the outcome of the data generating process which is of interest and assigning a numerical value to different possible outcomes. For example:

- the number of years (to the nearest whole number) of imprisonment imposed

- a 0 or 1 is assigned, depending on whether a device is faulty or not

- the mass of a puppy at four months

- the score on a die.

B Introducing a random variable which will represent this numerical value and assigning probabilities to each possible outcome (or a probability density function in the case of a continuous variable). For example:

- Y has a geometric distribution with

$$P(y \text{ years in prison}) = \frac{3}{4}\left(\frac{1}{4}\right)^y \quad (y = 0, 1, \dots)$$

- F takes values 0 or 1, with $P(0) = 0.0002$, $P(1) = 0.9998$

- M is normally distributed with mean 4.2 kg, standard deviation of 0.6 kg

- N takes values 1 to 6, each with probability $\frac{1}{6}$.

The aim is to ensure that the model is an adequate representation of those aspects of the data generating process in which you are interested. The information that enables you to do this is usually obtained from *sampling* the outcomes of the data generating process.

A *sampling process* is the way in which some of the outcomes are selected for consideration and recorded. For example:

- the selection of one week's convictions at each Crown Court in England, noting the gender of the defendant, the offence and the sentence

- the testing of every 100th device appearing from the production line and the recording of the result and time of testing

- the selection from a breeder's records of one puppy from each litter produced by her bitches over the past five years and its mass at four months

- the decision to roll a die 60 times and observe the score on each.

Before this information can be related to the model of the data-generating process, you must make a mathematical model of the sampling process. For complicated sampling processes, such as the 'puppy' example above or for stratified sampling, this is beyond the scope of this book, but there is one important special case which can be modelled reasonably simply.

If the outcome of the data-generating processes is modelled by the random variable X, then the process of taking an *independent random sample* of size n is modelled by the set of n random variables $X_1, X_2, \ldots, X_n$, each of which has a distribution identical to that of X, and all of which are independent of each other.

In the case of a finite population, for instance all the fish in a particular lake, when members of the population are not replaced after sampling, the random variables representing successive members of the sample are not strictly independent. However, if the size of the population is large then the independent random sampling model is still likely to be adequate for most purposes.

Sample statistics

You have already seen examples of the three main ways in which statisticians use the information in a sample to try to ensure that their model is an adequate representation of the data-generating process.

1 ESTIMATING A PARAMETER

Given that, when a greengrocer buys pears from a wholesaler he does not know the details of how some come to be bruised (the data-generating process), he could model this by a random variable which takes the values 1 (bruised) and 0 (unbruised) with constant probability. Then, given a sample of fruit, some of which is bruised, he could estimate this probability by finding the fraction of the sample which is bruised. The process of estimation will be considered in detail in Chapter 7.

2 TESTING AN HYPOTHESIS

Given that you do not know the details of what determines how many eggs are laid by a given jellyfish (the data-generating process), you could model this by a random variable which has a normal distribution for which the mean and standard deviation have been reliably estimated for a particular species. Then, given the numbers of eggs laid by a sample of jellyfish living on a polluted beach, you could test the hypothesis that the reproduction rate of these jellyfish is lower than average, by calculating the mean number of eggs per jellyfish in the sample.

Given that you do not know the details of what determines how many road accidents there will be on the M21 on a given day (the data-generating process), you could model this by a random variable which has a Poisson distribution. Then, given the numbers of accidents on a sample of days this year you could determine the χ^2 value which measures how well the data fit the claimed distribution.

In all these examples, some of the details of the sample were discarded, and only a single figure calculated from the sample was used by the statistical investigator. In **1**, no account was taken of which pears were bruised and only the total number of bruised pears was used. In **2**, the number of eggs laid by each jellyfish was known, but only the single figure of their mean was used. In **3**, the actual numbers of accidents on all the sample days were known but only the single χ^2 value was used to summarise them.

It is standard statistical practice to discard much of the information in the sample, either because it is not relevant to the aspect of the model being investigated or just in order to simplify the analysis. When the values of a random variable in a sample are used to calculate a single figure which summarises these values in some way, this figure is called a *sample statistic*.

Distributions of sample statistics

In more advanced statistical work, it is often essential to know the distribution, or at least the mean and perhaps variance, of a sample statistic, given the distribution of the random variable modelling the data-generating process. This section discusses two ideas which are often useful in this.

Distributions of functions of random variables

Suppose that the random variable X has the distribution

x	-2	-1	0	1	2
$P(X = x)$	0.08	0.24	0.36	0.24	0.08

What is the distribution of X^2?

The possible values of X^2 are the squares of the possible values of X, that is: 0, 1, 4.

The value of X^2 will be 0 precisely when the value of X is 0, so:

$$P(X^2 = 0) = P(X = 0) = 0.36$$

but the value of X^2 will be 1 when the value of X is 1 or -1, so:

$$P(X^2 = 1) = P(X = 1) + P(X = -1) = 0.48$$

and similarly with $P(X^2 = 4)$. Thus the complete distribution of X^2 is:

x	0	1	4
$P(X^2 = x)$	0.36	0.48	0.16

In general, if X is a random variable and $Y = g(X)$ is a function of X, then the probability that $g(X)$ takes a particular value y is the sum of the probabilities of all the x values with $g(x) = y$. Formally:

$$P(Y = y) = P(g(X) = y) = \sum_{g(x)=y} P(X = x).$$

The expected value of a function of X can be calculated in two ways. In the example:

$$E[X^2] = \sum_{y=0,1,4} y \times P(X^2 = y) = 0 \times 0.36 + 1 \times 0.48 + 4 \times 0.16 = 1.12$$

by definition of the expectation. However, this is the same thing as:

$$E[X^2] = \sum_{x=-2,-1,0,1,2} x^2 \times P(X = x)$$

$$= (-2)^2 \times 0.08 + (-1)^2 \times 0.24 + 0^2 \times 0.36 + 1^2 \times 0.24 + 2^2 \times 0.08$$

because, for instance, $(-2)^2 = 2^2 = 4$ and $P(X = -2) + P(X = 2) = P(X^2 = 4)$.

In general, if X is a random variable and $Y = g(X)$ is a function of X, then the expectation of $g(X)$ is, by definition:

$$E[g(X)] = \sum_{y} y \times P(g(X) = y)$$

Because of the result for $P(g(X) = y)$ above, however, this can be rewritten:

$$E[g(X)] = \sum_{y} y \times \sum_{g(x)=y} P(X = x) = \sum_{x} g(x) \times P(X = x).$$

This second form is often easier to calculate.

For a continuous random variable, the situation is rather less straightforward, as the distribution is given as a *probability density function* (p.d.f.), rather than a set of probabilities. In this case, the route from the p.d.f. of a random variable to the p.d.f. of a function of this variable is usually via the cumulative distribution function.

Suppose that the random variable X has the p.d.f.:

$$f(x) = \begin{cases} 0 & x < 1 \\ \dfrac{3}{x^4} & x \geqslant 1 \end{cases}$$

What is the p.d.f. of $Y = X^2$?

The cumulative distribution function of X is, by definition

$$F_X(x) = P(X \leqslant x) = \begin{cases} 0 & x < 1 \\ \int_1^x \dfrac{3}{t^4} \, dt & x \geqslant 1 \end{cases}$$

$$= \begin{cases} 0 & x < 1 \\ 1 - \dfrac{1}{x^3} & x \geqslant 1 \end{cases}$$

and the definition of the cumulative distribution function of X^2 is

$$F_{X^2}(x) = P(X^2 \leqslant x)$$

but X^2 is less than or equal to x whenever X is less than or equal to $\sqrt{x}$ (since $x > 0$) so that

$$F_{X^2}(x) = P(X \leqslant \sqrt{x})$$

$$= F_X(\sqrt{x})$$

$$= \begin{cases} 0 & x < 1 \\ 1 - x^{-\frac{3}{2}} & x \geqslant 1 \end{cases}$$

Finally, use the relationship

$$f(x) = \frac{d}{dx} F(x)$$

between the density function and cumulative distribution function to give the density function of X^2

$$f_{X^2}(x) = \begin{cases} 0 & x < 1 \\ \dfrac{3}{2} x^{-\frac{5}{2}} & x \geqslant 1 \end{cases}$$

The expectation of X^2 can be calculated, as in the discrete case, in two ways.

First, by definition:

$$E[X^2] = \int x f_{X^2}(x) \, dx$$

$$= \int_1^\infty \frac{3}{2} x^{-\frac{3}{2}} \, dx$$

$$= \left[-3x^{-\frac{1}{2}} \right]_1^\infty = 0 - (-3) = 3$$

But the same result is obtained from the integral:

$$E[X^2] = \int x^2 f_X(x) \, dx$$

$$= \int_1^\infty 3x^{-2} \, dx$$

$$= \left[-3x^{-1} \right]_1^\infty = 0 - (-3) = 3$$

In general, if X is a continuous random variable and $Y = g(X)$ is a function of X, then the expectation of $g(X)$ is, by definition:

$$E[g(X)] = \int y f_{g(X)}(y) \, dy$$

where $f_{g(X)}(y)$ is the density function of $g(X)$.

However, this can be rewritten:

$$E[g(X)] = \int g(x) f_X(x) \, dx$$

using the density function, $f_X(x)$, of X itself. Again, this second form is often easier to calculate.

The formulae for the variance of a function of a random variable follow automatically from those for the expectation because, by definition, $\text{Var}[X] = E[X - E[X]^2] = E[X^2] - E[X]^2$.

For a discrete variable, then:

$$\text{Var}[g(X)] = \sum_x \{g(x) - E[g(X)]\}^2 P(X = x) = \sum_x \{g(x)\}^2 P(X = x) - E[g(X)]^2$$

while in the continuous case:

$$\text{Var}[g(X)] = \int \{g(x) - E[g(X)]\}^2 f_X(x) \, dx = \int \{g(x)\}^2 f_X(x) \, dx - E[g(X)]^2$$

EXAMPLE 2.1

The distribution of magnitudes M of a particular type of star is given by the density function:

$$f(m) = \frac{1}{4} m e^{-\frac{m}{2}}$$

The magnitude of a star is related to its luminosity (brightness), L, by:

$$L = 10^{-\frac{2}{5}m} = e^{-\frac{2}{5}m \ln 10}$$

Find the expected brightness of this type of star and the variance of the brightness.

SOLUTION

Use the result $\int_0^\infty t e^{-at} \, dt = a^{-2}$.

$$E[L] = \int L(m) f(m) \, dm$$

$$= \int_0^\infty e^{-\frac{2}{5}m \ln 10} \cdot \frac{1}{4} m e^{-\frac{1}{2}m} \, dm$$

$$= \frac{1}{4} \int_0^\infty m e^{-\left(\frac{2}{5} \ln 10 + \frac{1}{2}\right)m} \, dm$$

$$= \frac{1}{4} \left(\frac{2}{5} \ln 10 + \frac{1}{2}\right)^{-2} \approx 0.124$$

$$\text{Var}[L] = \int \{L(m)\}^2 f(m) \, dm - \{E[L]\}^2$$

$$= \int_0^\infty \left\{ e^{-\frac{2}{5}m \ln 10} \right\}^2 \cdot \frac{1}{4} me^{-\frac{1}{2}m} \, dm - \{E[L]\}^2$$

$$= \frac{1}{4} \int_0^\infty me^{-\left(\frac{4}{5}\ln 10 + \frac{1}{2}\right)m} \, dm - \{E[L]\}^2$$

$$= \frac{1}{4} \left\{ \frac{4}{5}\ln 10 + \frac{1}{2} \right\}^{-2} - \{E[L]\}^2 \approx 0.0302$$

EXAMPLE 2.2

When R users are logged on to the Internet via a particular server, the average speed (in kilobytes per second) at which each person's data are transferred is $T = 30\left(\frac{8}{9}\right)^R$.

If the number of users logged on at any one time has a binomial distribution with $n = 20$ and $p = \frac{1}{4}$, find the expectation and variance of the average speed at which data is transferred.

SOLUTION

Use the result (binomial expansion) $\displaystyle\sum_{r=0}^n {}^nC_r p^r q^{n-r} = (p+q)^n$.

$$E[T] = \sum_r T(r)P(R = r) = \sum_{r=0}^{20} 30\left(\frac{8}{9}\right)^r P(R = r)$$

$$= \sum_{r=0}^{20} 30\left(\frac{8}{9}\right)^r \binom{20}{r} \left(\frac{1}{4}\right)^r \left(\frac{3}{4}\right)^{20-r}$$

$$= 30\sum_{r=0}^{20} \binom{20}{r} \left(\frac{8}{9} \times \frac{1}{4}\right)^r \left(\frac{3}{4}\right)^{20-r} = 30\sum_{r=0}^{20} \binom{20}{r} \left(\frac{2}{9}\right)^r \left(\frac{3}{4}\right)^{20-r}$$

$$= 30\left(\frac{2}{9} + \frac{3}{4}\right)^{20} = 30\left(\frac{35}{36}\right)^{20} = 17.08$$

$$E[T^2] = \sum_r \{T(r)\}^2 P(R = r) = \sum_{r=0}^{20} \left\{30\left(\frac{8}{9}\right)^r\right\}^2 P(R = r)$$

$$= \sum_{r=0}^{20} 900\left(\left(\frac{8}{9}\right)^2\right)^r \binom{20}{r} \left(\frac{1}{4}\right)^r \left(\frac{3}{4}\right)^{20-r}$$

$$= 900\sum_{r=0}^{20} \binom{20}{r} \left(\frac{64}{81} \times \frac{1}{4}\right)^r \left(\frac{3}{4}\right)^{20-r} = 900\sum_{r=0}^{20} \binom{20}{r} \left(\frac{16}{81}\right)^r \left(\frac{3}{4}\right)^{20-r}$$

$$= 900\left(\frac{16}{81} + \frac{3}{4}\right)^{20} = 900\left(\frac{307}{324}\right)^{20} = 306.27$$

so $\text{Var}[T] = 306.27 - (17.08)^2 = 14.54$

Distributions of combinations of random variables

You have modelled a sample as n independent random variables, with identical distributions. You already know some results about distributions of combinations of such sets of random variables, which will be revised and extended here.

Distributions of linear combinations of random variables

A *linear combination* of a set of n random variables X_i is a sum

$$a_1X_1 + a_2X_2 + \ldots + a_nX_n = \sum_{i=1}^{n} a_iX_i$$

where the a_i are any set of constants.

A In general, the distribution of a linear combination of random variables is not simply related to the distribution of the variables being combined, but you saw in *Statistics 3* that, if the X_i are all normal, the linear combination will also be normal.

B It is true for any linear combination of random variables that:

$$E[a_1X_1 + a_2X_2 + \ldots + a_nX_n] = a_1E[X_1] + a_2E[X_2] + \ldots + a_nE[X_n]$$

$$\text{or} \quad E\left[\sum_{i=1}^{n} a_iX_i\right] = \sum_{i=1}^{n} a_iE[X_i]$$

If the random variables are identically distributed, then each X_i has the same expectation, say $E[X_i] = \mu$, and then:

$$E\left[\sum_{i=1}^{n} a_iX_i\right] = \mu\left(\sum_{i=1}^{n} a_i\right)$$

C It is true for any linear combination of *independent* random variables that:

$$\text{Var}[a_1X_1 + a_2X_2 + \ldots + a_nX_n] = a_1^2\text{Var}[X_1] + a_2^2\text{Var}[X_2] + \ldots + a_n^2\text{Var}[X_n]$$

$$\text{or} \quad \text{Var}\left[\sum_{i=1}^{n} a_i^2X_i\right] = \sum_{i=1}^{n} a_i^2\text{Var}[X_i]$$

If the random variables are identically distributed, then each X_i has the same variance, say $E[X_i] = \sigma^2$, and then:

$$\text{Var}\left[\sum_{i=1}^{n} a_iX_i\right] = \sigma^2\left(\sum_{i=1}^{n} a_i^2\right)$$

D An important special case is the situation where each $a_i = \dfrac{1}{n}$. Then we define the combination:

$$\overline{X} = \frac{X_1 + X_2 + \ldots + X_n}{n} = \frac{1}{n}\sum_{i=1}^{n} X_i$$

called the *sample mean*. In this case, if the X_i are identically distributed and independent, then the results in **B** and **C** above reduce to:

$$\mathrm{E}[\overline{X}] = \mu, \quad \mathrm{Var}[\overline{X}] = \frac{\sigma^2}{n}.$$

EXAMPLE 2.3

When a 'ready-meal' curry is being assembled, the number of grams of chicken per portion is normally distributed with mean 80 and variance 230. The number of grams of sauce added is also normally distributed with mean 55 and variance 140 and is independent of the amount of chicken added. The portion is then made up to exactly 350 grams with rice.

Chicken contains 1.8 kilocalories per gram, the sauce 4.5 kilocalories per gram and the rice 2.6 kilocalories per gram. What is the distribution of the random variable K which gives the calorific value of the portion? If the testing process involves determining the mean calorific value of random samples of twelve portions, what will be the distribution of this sample mean?

SOLUTION

The variables are $C \sim N(80, 230)$, for the weight of chicken and $S \sim N(55, 140)$ for the weight of sauce. Then:

$$K = 1.8C + 4.5S + 2.6(350 - C - S) = 910 - 0.8C + 1.9S$$

K is normally distributed, because C and S are, and:

$$\mathrm{E}[K] = 910 - 0.8\mathrm{E}[C] + 1.9\mathrm{E}[S] = 950.5$$

$$\mathrm{Var}[K] = (-0.8)^2\mathrm{Var}[C] + (1.9)^2\mathrm{Var}[S] = 652.6$$

The variable $\overline{K}$, which is the mean calorific value of a sample of twelve portions, is also normally distributed with:

$$\mathrm{E}[\overline{K}] = \mathrm{E}[K] = 950.5$$

$$\mathrm{Var}[\overline{K}] = \frac{\mathrm{Var}[K]}{12} = 54.38$$

Maximum and minimum

Given a set of n random variables $X_1, X_2, \ldots, X_n$ constituting a sample, the largest of the n values taken, is itself a random variable, the *sample maximum L*.

When X is a continuous variable, the distribution of L can be found by noting that:

- if all n of a set of numbers are less than the value ℓ, then the largest of them is also less than ℓ

- if the largest of a set of n numbers is less than a value ℓ, then all of them are less than ℓ.

Thus, the states of affairs $\{L \leqslant \ell\}$ and $\{X_1 \text{ and } X_2 \text{ and } \ldots \text{ and } X_n \leqslant \ell\}$ are identical.

So, because the X_i are independent:

$$P(L \leqslant \ell) = P(X_1 \text{ and } X_2 \text{ and} \ldots \text{and } X_n \leqslant \ell)$$
$$= P(X_1 \leqslant \ell)P(X_2 \leqslant \ell) \ldots P(X_n \leqslant \ell)$$

and because $X_1, X_2, \ldots, X_n$ are identically distributed, with the same distribution as the random variable X which models the data-generating process, this reduces to

$$P(L \leqslant \ell) = \{P(X \leqslant \ell)\}^n.$$

But the left-hand side of this is just, by definition, the cumulative distribution function for L and the right-hand side can similarly be written in terms of the cumulative distribution function of X

$$F_L(\ell) = \{F_X(\ell)\}^n.$$

Finally, the density function for L is obtained by differentiating the cumulative distribution function.

EXAMPLE 2.4

The time T (in minutes) for which a passenger must wait for a train each morning has an exponential distribution with mean 8, that is:

$$f_T(x) = \frac{1}{8}e^{-\frac{1}{8}x} \quad (0 \leqslant x < \infty)$$

Find the distribution of the maximum wait in a sample of three mornings, and determine its mean.

SOLUTION

The cumulative distribution function of T is:

$$F_T(x) = \int_0^x \frac{1}{8}e^{-\frac{1}{8}t} \, dt = 1 - e^{-\frac{1}{8}x}$$

so the cumulative distribution function of L, the longest of the three waits, is:

$$F_L(x) = \left(1 - e^{-\frac{1}{8}x}\right)^3$$

and so its density function is:

$$f_L(x) = \frac{d}{dx}F_L(x) = \frac{3}{8}e^{-\frac{1}{8}x}\left(1 - e^{-\frac{1}{8}x}\right)^2 = \frac{3}{8}\left(e^{-\frac{1}{8}x} - 2e^{-\frac{1}{4}x} + e^{-\frac{3}{8}x}\right)$$

The expectation can then be calculated, using the result $\int_0^\infty xe^{-ax}\,dx = a^{-2}$, as

$$E[L] = \int_0^\infty xf_L(x)\,dx = \int_0^\infty \frac{3}{8}x\left(e^{-\frac{1}{8}x} - 2e^{-\frac{1}{4}x} + e^{-\frac{3}{8}x}\right)\,dx$$

$$= \frac{3}{8}\left(64 - 2 \times 16 + \frac{64}{9}\right) = \frac{44}{3}$$

A similar technique is applicable in the case of discrete distributions, but here the formal method is unnecessarily complicated. It is easier to understand the idea from examples.

EXAMPLE 2.5

The probability of winning in a one-armed bandit game is $\frac{1}{4}$. This means that the number W of wins in five games has a binomial distribution with:

$$P(W = w) = \binom{5}{w}\left(\frac{1}{4}\right)^w\left(\frac{3}{4}\right)^{5-w} \qquad (0 \leqslant w \leqslant 5)$$

If seven people play the game, find the distribution of the maximum number of wins any of them achieves and the expectation of this number.

SOLUTION

The table shows the probability of each possible value of W and the cumulative distribution function for this variable.

w	0	1	2	3	4	5
$P(W = w)$	0.2373	0.3955	0.2637	0.0879	0.0146	0.0010
$P(W \leqslant w)$	0.2373	0.6328	0.8965	0.9844	0.9990	1.0000

As in the continuous case, the probability that the greatest of seven numbers is less than or equal to m is just the probability that each of them is less than or equal to m. So, if M is the maximum number of wins, you can construct the cumulative distribution function for M as

$$P(M \leqslant m) = (P(W \leqslant m))^7.$$

m	0	1	2	3	4	5
$P(M = m)$	0.0000	0.0406	0.4654	0.8956	0.9932	1.0000

Finally, notice, for example, that the difference between '3 or fewer wins' and '2 or fewer wins' is 'exactly 3 wins'. In general

$$P(M = m) = P(M \leqslant m) - P(M \leqslant m - 1)$$

so that the distribution of M is:

m	0	1	2	3	4	5
$P(M = m)$	0.0000	0.0406	0.4248	0.4302	0.0976	0.0068

and its expectation is:

$$E[M] = \sum_{m=0}^{5} m \times P(M = m) = 2.605.$$

EXAMPLE 2.6

The probability of winning in a scratchcard game is $\frac{1}{12}$.

This means that the number S of scratchcards bought before the first win has a geometric distribution with:

$$P(S = s) = \frac{1}{12}\left(\frac{11}{12}\right)^{s-1} \qquad (1 \leqslant s < \infty)$$

If three people buy scratchcards, find the distribution of the number of scratchcards bought before they have all won at least once and the expectation of this number.

SOLUTION

The cumulative distribution function of S is:

$$P(S \leqslant s) = \sum_{i=1}^{s} \frac{1}{12}\left(\frac{11}{12}\right)^{i-1} = 1 - \left(\frac{11}{12}\right)^{s}$$

Again, the probability that the longest of three waits is less than or equal to s is just the probability that each of the waits is less than or equal to s. So, if L is the longest wait:

$$P(L \leqslant \ell) = \left(1 - \left(\frac{11}{12}\right)^{\ell}\right)^{3}$$

You can use the same method as before to find the distribution of L, because:

$$P(L = \ell) = P(L \leqslant \ell) - P(L \leqslant \ell - 1)$$

$$= \left(1 - \left(\frac{11}{12}\right)^{\ell}\right)^{3} - \left(1 - \left(\frac{11}{12}\right)^{\ell-1}\right)^{3}$$

$$= 3\left(1 - \left(\frac{11}{12}\right)\right)\left(\frac{11}{12}\right)^{\ell-1} - 3\left(1 - \left(\frac{11}{12}\right)^{2}\right)\left(\left(\frac{11}{12}\right)^{2}\right)^{\ell-1}$$

$$+ \left(1 - \left(\frac{11}{12}\right)^{3}\right)\left(\left(\frac{11}{12}\right)^{3}\right)^{\ell-1}$$

A graph of this probability distribution is shown in Figure 2.1.

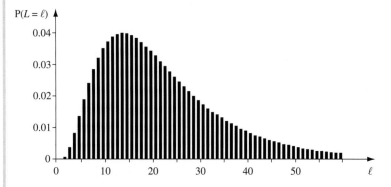

Figure 2.1

The expectation of L is, by definition

$$E[L] = \sum_{\ell=1}^{\infty} \ell \times P(L = \ell)$$

$$= 3 \sum_{\ell=1}^{\infty} \ell \left(1 - \left(\frac{11}{12} \right) \right) \left(\frac{11}{12} \right)^{\ell-1} - 3 \sum_{\ell=1}^{\infty} \ell \left(1 - \left(\frac{11}{12} \right)^2 \right) \left(\left(\frac{11}{12} \right)^2 \right)^{\ell-1}$$

$$+ \sum_{\ell=1}^{\infty} \ell \left(1 - \left(\frac{11}{12} \right)^3 \right) \left(\left(\frac{11}{12} \right)^3 \right)^{\ell-1}$$

which can be evaluated using the result

$$\sum_{\ell=1}^{\infty} \ell (1 - p) p^{\ell-1} = \frac{1}{1-p}$$

to give:

$$E[L] = \frac{3}{1 - \left(\dfrac{11}{12} \right)} - \frac{3}{1 - \left(\dfrac{11}{12} \right)^2} + \frac{1}{1 - \left(\dfrac{11}{12} \right)^3} = 21.57.$$

ACTIVITY

1 Decide how to find the distribution of the least value taken by a random variable in a sample.

Apply your method to Examples 2.3, 2.4 and 2.5.

2 Look back at the work you have done in earlier Statistics components. Discuss how this fits into the general pattern of statistical analysis outlined above.

1 The random variable X has a rectangular distribution on $[0, 1]$.

(i) Find the mean and variance of the variable $P = 4\sqrt{1 - X^2}$.

(ii) If $X_1, X_2, \ldots, X_n$ are a sample of n random numbers on $[0, 1]$, find the mean and variance of

$$\overline{P} = \frac{1}{n}(P_1 + P_2 + \ldots + P_n)$$

in terms of n.

(iii) Explain how to use a sample of n random numbers on $[0, 1]$ to estimate π, and calculate how large n would have to be before you could expect accuracy to 3 decimal places.

2 A jar contains a large number of mixed black and white peppercorns. When ten of these are removed at random and the number W of white peppercorns removed is counted, you might use $\frac{W}{10}$ as an estimate of the proportion p of white peppercorns in the jar. If the jar is large enough to ignore the fact that you are sampling without replacement, W will have an approximately binomial distribution $B(10, p)$ with variance $10p(1 - p)$, so you might use:

$$S = \sqrt{10\frac{W}{10}\left(1 - \frac{W}{10}\right)} = \sqrt{\frac{W(10 - W)}{10}}$$

as an estimate of the standard deviation of W. Find the distribution of S if $p = \frac{1}{3}$ and find its mean and variance.

3 The dyadic size $|n|_2$ of an integer n is the reciprocal of the largest power of 2 that divides the number. For instance, $|56|_2 = \frac{1}{8}$, because 8 is the highest power of 2 to divide 56, and $|55|_2 = 1$, because $1 = 2^0$ is the highest power of 2 to divide 55. If N is a random variable uniformly distributed on the integers from 1 to 100, find the distribution of $|N|_2$. Find also the mean and variance of $|N|_2$.

If N is instead uniformly distributed on the integers from 1 to a large number L, investigate what happens to the mean of $|N|_2$ as $L \to \infty$.

4 When a ball is thrown vertically upwards from ground level, the height H (in metres) that it reaches is a function:

$$H = \frac{1}{20}U^2$$

of the speed U (in metres per second) with which it is thrown.

The ball is thrown in such a way that U has a distribution with probability density function:

$$f(x) = \frac{1}{4}\left(1 - \left(\frac{x - 20}{3}\right)^2\right) \quad (17 \leqslant x \leqslant 23)$$

(i) Find the density function for the variable L which is the larger of the two initial speeds of two balls thrown independently in this way. Find the mean of L.

(ii) Find the mean and variance of H.

(iii) When a sample of five balls are thrown independently in this way, the mean of their heights is M. Find the mean and variance of M.

5 (i) Find the expectation and variance of the score on a single tetrahedral die, with sides numbered 1, 2, 3, 4.

Three such dice are thrown.

(ii) Find the distribution of the largest score obtained and calculate its expectation and variance.

(iii) Find the distribution of the sum of the three scores.

Hence calculate explicitly the expectation and variance of the mean of the three scores, and show that they are related as you expect to your answers to **(i)**.

6 An A-level component has a written paper marked out of 70 and a coursework task marked out of 14. The final mark must be out of 60, with the written paper counting for 80% and the coursework task 20%. For a particular candidate, the written mark is modelled as a normal random variable $W \sim N(46, 49)$ and the coursework mark is modelled as an independent normal random variable $C \sim N(9, 4)$.

(i) Explain why the variable T which records the candidate's final mark out of 60 is given by:

$$T = \frac{24}{35}W + \frac{6}{7}C$$

(ii) Find the mean and variance of T, and hence find the probability that the candidate gains an A grade, if she must achieve a final mark of 47 or more out of 60 to do this.

7 In a game, each player throws four ordinary six-sided dice. The random variable X is the largest number showing on the dice.

(i) Find the probability that $X = 1$.

(ii) Find the probability that $X \leqslant 2$ and deduce that the probability that $X = 2$ is $\frac{5}{432}$.

(iii) Find the probability that $X = 3$.

(iv) Find the probability that $X = 6$, and explain without further calculation why 6 is the most likely value of X.

8 The continuous random variable X has a rectangular (or uniform) distribution on the interval $0 \leqslant x \leqslant 4$.

(i) Write down the probability density function for X, and sketch its graph.

(ii) Find the cumulative distribution function $F(x) = P(X \leqslant x)$.

A second random variable, Y, is related to X by the equation $Y = \sqrt{X}$.

(iii) Show that $P(Y \leqslant y) = P(X \leqslant y^2) = \frac{1}{4}y^2$.

Using this, the cumulative distribution function for Y, find the probability density function for Y. State the range of values which Y takes.

(iv) Show that the median value of Y is the square root of the median value of X.

9 A secretary types letters on to sheets of paper 30 cm long and then folds the letters as shown.

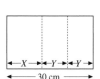

The first fold is X cm from one edge of the paper and the second, Y cm from the other edge, is exactly in the middle of the remainder of the paper, so that

$$Y = \frac{1}{2}(30 - X).$$

The distance X cm is normally distributed with mean 10.2 cm and standard deviation 1.2 cm.

(i) Obtain the distribution of Y.

(ii) The letters have to fit into envelopes 11 cm wide. Find

(a) $P(X > 11)$

(b) $P(Y > 11)$

(c) the proportion of folded letters which *will* fit into the envelope.

(iii) The 'overlap' is $X - Y$. Show that

$X - Y = \dfrac{3}{2}(X - 10)$ and hence verify that

Var$(X - Y)$ is *not* equal to

Var(X) + Var(Y).

(iv) Explain why the rule

Var$(aX + bY) = a^2$Var$(X) + b^2$Var(Y)

does not apply in this case.

10 For a positive integer r, the rth moment of the distribution of the random variable X about its mean μ is denoted by μ_r and defined by

$$\mu_r = E[(X - \mu)^r].$$

(i) Show that $\mu_1 = 0$ for any X.

(ii) What is the familiar interpretation of μ_2?

(iii) μ_3 is often interpreted as a measure of the *skewness* of a distribution. What will be the value of μ_3 for a symmetrical distribution?

(iv) Prove the results

(a) $\mu_2 = E[X^2] - (E[X])^2$

(b) $\mu_3 = E[X^3] - 3E[X]E[X^2] + 2(E[X])^3$.

(v) Using the results of **(iv)**, or otherwise, find the values of μ_2 and μ_3 for the continuous random variable Y with probability density function

$$f(y) = \begin{cases} 3y^2 & 0 < y < 1 \\ 0 & \text{elsewhere} \end{cases}$$

State whether this is a symmetrical distribution.

(vi) The discrete random variable T has

$$P(T = 3) = \frac{1}{10}, P(T = 1) = \frac{1}{2},$$

$$P(T = -2) = \frac{2}{5}.$$

Find the value of μ_3 for T. Is this a symmetrical distribution?

1 If a random variable Y is a function $Y = g(X)$ of a random variable X then

- if X, Y are discrete

$$E[Y] = \sum_x g(x)P(X = x)$$

$$Var[Y] = \sum_x (g(x))^2 P(X = x) - E[Y]^2$$

- if X, Y are continuous, and X has density function $f(x)$

$$E[Y] = \int g(x)f(x)\, dx$$

$$Var[Y] = \int (g(x))^2 f(x)\, dx - E[Y]^2$$

2 If a random variable Y is a linear combination $Y = \sum_{i=1}^{n} a_i X_i$ of the n random variables X_i then

- if the X_i are normally distributed then so is Y

- in general

$$E[Y] = \sum_{i=1}^{n} a_i E[X_i]$$

- if the X_i are independent then

$$Var[Y] = \sum_{i=1}^{n} a_i^2 Var[X_i]$$

- in the case where the X_i are identically distributed with mean μ and variance σ^2, the specific linear combination $\overline{X} = \dfrac{1}{n}\sum_{i=1}^{n} X_i$, has

$$E[\overline{X}] = \mu, \quad Var[\overline{X}] = \frac{\sigma^2}{n}$$

3 If a random variable Y is the largest of the n random variables X_i, which are identically distributed with cumulative distribution function $F_X(x)$, then the cumulative distribution function of Y is

$$F_Y(x) = (F_X(x))^n$$

3 Hypothesis tests for the difference of means

Let not thy left hand know what thy right hand doeth.

St Matthew 6:3

When he returned to the office at two o'clock, his desk was clear. Until the afternoon post arrived, he had nothing to do. A ray of sunlight shone through a high window and he watched the specks of dust fall gently through it. With one finger, he pushed a paperclip around his blotter. The only sound was the distant tapping of a typewriter in a room down the corridor. For what seemed like hours he sat, sunk in a deep reverie. Then he lowered his eyes and glanced at his watch. It was five past two.

Have you noticed how time often seems to pass more slowly after lunch?

If time passes more slowly, one minute of real time should seem longer, so if you ask people to estimate when a minute appears to have elapsed, the real time elapsed will be less.

You could ask the question: 'Will the mean real time elapsed when one minute appears to have elapsed be less after lunch than before?'

In this example you are interested, not in what the mean value of a random variable is, but in what the difference between the mean values is in two different situations. Statistical problems giving rise to different versions of this general question are the topic of this chapter.

Find a group of volunteers. Give each of them a starting signal and ask them to say when one minute has elapsed. Record the real time elapsed. You will need to ask each person twice: once before lunch and once after lunch.

Consider the following questions before you start.

- Why should your volunteers not know the purpose of the experiment?

- Should each volunteer be tested before and after lunch on the same day?

- Would it be better to test all your subjects on the same day or spread them throughout the week?

The next few sections will be illustrated in the text by the data below, from an experiment carried out as above on twelve volunteers. If you prefer, you can use the data you have collected yourself to follow the work.

Notice how most people seem to underestimate, both before and after lunch.

Initials of subject	PH	GG	RS	TP	CF	GJ	ES	RP	CN	IH	DS	RB
Before-lunch estimate (seconds)	53	44	55	63	44	59	54	52	50	36	62	43
After-lunch estimate (seconds)	51	37	59	48	39	48	57	40	45	33	63	51

The volunteers in research projects are called *subjects* and 'before lunch' and 'after lunch' are the two *conditions* in which they were tested.

An experiment where one set of subjects is tested in each of two conditions is called a *paired design*, whereas if separate sets of subjects are tested in the two conditions this is an *unpaired design*.

The paired sample *t*-test

You are testing the hypotheses:

H_0: There is, on average, no difference between people's estimates of one minute before and after lunch.

H_1: After lunch, people's estimates of one minute tend to be shorter.

The form of the alternative hypothesis, which claims that estimates are shorter rather than just different, indicates that a one-tailed test is appropriate. This is used because the belief that motivated our test in the first place was the one-sided suggestion that time passed more slowly after lunch.

Let D be the random variable 'difference between before- and after-lunch times'. That is, the value of D is the result of subtracting the estimate after lunch from the estimate before lunch for a randomly chosen member of the population. The measurements made on the sample of size twelve correspond to values of the variables D_i $(i = 1, 2, \ldots, 12)$, each of which has the distribution of D. These values are referred to with the notation d_i $(i = 1, 2, \ldots, 12)$.

Value of i	1	2	3	4	5	6	7	8	9	10	11	12
Initials of subject	PH	GG	RS	TP	CF	GJ	ES	RP	CN	IH	DS	RB
Result d_i of subtracting estimate after lunch from estimate before lunch	2	7	−4	15	5	11	−3	12	5	3	−1	−8

If these figures came from a population where D had a mean of zero, then you would expect to see approximately equal numbers and sizes of positive and negative differences in the sample. What you are trying to do is to decide whether the twelve differences listed above are, on the whole, so obviously positive that it is unlikely that they came from a population where D had a mean of zero.

A sensible measure of the average tendency in the sample for these differences to be positive is the mean of the sample differences, which takes the value $\overline{d} = \dfrac{11}{3}$.

In line with the usual hypothesis testing principles, you need to discover whether such a result is reasonably likely to arise by chance from a sample of size twelve when the null hypothesis is true. To do this you need to know the sampling distribution of the statistic $\overline{D}$, the mean of the D_i.

To find this distribution, assume that D is a normal random variable: the null hypothesis then says that this random variable has mean zero. From *Statistics 3*, if D is a normal random variable with mean zero and S^2 is the unbiased estimator of its variance from a sample of size n, then the sampling distribution of the statistic:

$$T = \frac{\overline{D}}{\dfrac{S}{\sqrt{n}}}$$

is a t-distribution with $n - 1$ degrees of freedom. In our case $n = 12$, so the test statistic has 11 degrees of freedom and $s = 6.946$, so that the denominator of the test statistic, which is referred to as the standard error of the mean, is:

$$\frac{s}{\sqrt{n}} = 2.005$$

and the test statistic itself is:

$$t = \frac{\overline{d}}{\dfrac{s}{\sqrt{n}}} = \frac{\dfrac{11}{3}}{2.005} = 1.829$$

The critical region, at the 5% significance level, for a one-tailed *t*-test with 11 degrees of freedom, is found in the *t*-tables under $n = 11$ and $p = 10\%$ (because the tables are constructed to give each tail a probability of $\frac{1}{2}p\%$). This gives a critical region of $t > 1.796$ which means that values of the test statistic larger than 1.796 would be unlikely (with probability less than 0.05) to arise by chance if the null hypothesis were true.

Since $1.829 > 1.796$, you reject the null hypothesis, and accept the alternative hypothesis that estimates are smaller in the afternoon.

Notice that you have done nothing different here from the *t*-testing in *Statistics 3*. The new feature is that the variable used in the test statistic was obtained as the difference of the two variables which were actually measured in the sample. This means that the conclusions you drew were about the mean difference between the two measured variables. The test carried out here is called a *paired sample t-test for the difference of two means*.

It is important to recognise the difference between a random variable and the value it takes in a particular sample. You use capital letters for random variables and the corresponding lower-case letters to indicate the values taken by these variables in a particular sample. For instance, the random variables in the example above are D_i ($i = 1$ to 12), $\overline{D}$, S and T, with values in the sample of d_i ($i = 1$ to 12), $\overline{d}$, s and t, respectively.

EXPERIMENTS

Test at least one of the hypotheses below by carrying out the experiment suggested. Each of them uses a paired sample *t*-test.

1 People can hold their breath for longer after deep breathing exercises. The two conditions are:

 (a) Take a deep breath and hold it for as long as possible.

 (b) Breathe in and out very deeply and slowly four times. On the fifth intake of breath, hold it for as long as possible.

2 People's reaction times are shorter first thing in the morning than late in the afternoon.

 You can use a 30 cm ruler as a reaction timer: hold the ruler vertically with the zero mark downwards, while the subject holds his thumb and forefinger 2 cm apart at the zero mark of the ruler. You drop the ruler without warning and your subject tries to catch it between thumb and forefinger. The distance, d, in millimetres, through which the ruler has fallen before it is caught can be used to measure the reaction time, t, in seconds, using the formula:

$$t = \frac{\sqrt{d}}{70}$$

You could also use the reaction timer to test whether people's performance improves with practice: give each subject a preliminary test, ten practice drops and then a final test and record the difference between the first and last reaction times.

3 Presenting words in groups with related meanings makes them easier to recall than presenting them in alphabetical order.

You will need to produce lists of words (about 25 words each works well) some organised by meaning, others alphabetically.

For example:

peach	lion	table	flute	beer
pear	puma	chair	violin	cola
apple	zebra	bed	cello	wine
grape	badger	sofa	drum	cider
apricot	elk	stool	piano	water

is organised by meaning, whereas:

apple	cello	elk	pear	table
apricot	chair	flute	piano	violin
badger	cider	grape	puma	water
bed	cola	lion	sofa	wine
beer	drum	peach	stool	zebra

is organised alphabetically.

Each subject should be given a list of words to study, without writing anything, for one minute. At the end of this time, ask your subject to count out loud backwards, going down in threes from 587 (the point of this is to clear the short-term memory). After about 30 seconds, give each subject a sheet of paper and ask them to write down as many words as they can remember in one minute. Each subject needs to be tested twice, once with a list in alphabetical order, once with a list organised by meaning – but not a list with the same words on, of course. It is a good idea to have a number of different lists of each type, so that if some words are simply more memorable than others (what if one list had 'hippopotamus' on?), this will not spoil the effect you are looking for. It is also sensible to give half the subjects the alphabetical list first, and half the list organised by meaning first, in case there is a 'practice effect'.

Assumptions for the *t*-test

There are two crucial assumptions in carrying out a *t*-test.

1 That the sample is random.

2 That the variable is normally distributed.

Was your sample random?

Strictly, this requires every possible sample to have an equal probability of being chosen. If you simply picked a group of volunteers, therefore, yours was probably not a random sample. However, this method is very close to the method often used by academic psychologists when choosing their samples. The hope in choosing a random sample is that the effects of all the irrelevant differences between members of the population which influence the variables you are testing will average out.

Many psychology experiments are done on samples of American university students: these are clearly not strictly random samples of the human population. However, the psychologist's argument is that the characteristics they are trying to measure are universal; that is, they are shared by everyone, independently of other characteristics such as age and level of education. This means that a random sample of American students will have the same properties as a random sample of the whole population. This argument is not accepted by everyone: some results verified on American students have later been found not to apply to groups of older, less well-educated or culturally different subjects.

Do you think that the effect you have been looking at is sufficiently universal to make deductions from your sample valid?

Are the differences between before- and after-lunch times normally distributed?

This is a plausible assumption, but not obviously correct. If you were doing this hypothesis test in earnest, you might want to make some prior check that normality of the differences was an acceptable assumption: there are various tests which can be used and you have met one of them in Chapter 1.

Notice that no claim is made about the distributions of before- and after-lunch times separately, but only about the distribution of differences. It is not necessary, for the *t*-test to be used, that the before- and after-lunch times themselves be normally distributed, nor even sufficient, since the times in the two conditions are unlikely to be independent.

 Go back to any results you have for the experiments suggested on pages 51–52. Do you think the two assumptions are justified for the experiments you undertook?

The underlying logic of hypothesis testing

When you construct the sampling distribution of a test statistic you use:

1 a model for the distribution of the random variables involved in the statistic

2 the value given to a parameter of this distribution by the null hypothesis.

In the time-estimation example, the construction of the sampling distribution depends on:

1 the differences between people's estimates of one minute before and after lunch being independent and distributed normally, with a common mean and variance

2 the null hypothesis that the mean of these differences is zero.

The alternative hypothesis, that the mean of these differences is greater than zero, gives an alternative range of possible values for the parameter of the distribution, but assumes the same model for the random variables involved.

In the example it was determined (by using pre-calculated tables, in fact), that if:

- the model for the random variables was correct

- the null hypothesis were true

then a test statistic greater than 1.796 would only arise in a random sample 5% of the time.

You obtained a value greater than 1.796, for which there are three possible explanations.

EXPLANATION A

1 The model is correct.

2 The null hypothesis is false, because the mean difference in before- and after-lunch times is greater than zero.

EXPLANATION B

1 The model is correct.

2 The null hypothesis is true (or false because the mean difference in before- and after-lunch times is actually less than zero).

However the sample selected happens to give a value of the test statistic greater than 1.796. The probability of this happening is 0.05 (the significance level) if the null hypothesis is true, or less if the mean difference in before- and after-lunch times is actually less than zero.

EXPLANATION C

1 The model is incorrect, because the sampling method does not produce independent differences for each subject, or because the differences are not distributed normally in the population, or do not have a common mean or variance.

2 The null hypothesis is true or false.

In this case you have no idea how likely it is that the test statistic will have any value at all.

The hypothesis testing methodology is

- to assume that explanation C is not the case

- to observe that if explanation B was the case then the results obtained would be very unlikely

- and therefore to accept that explanation A is the case.

Thus you reject the null hypothesis and accept the alternative.

However, you should always be aware that the logic which leads you to this conclusion on the basis of the evidence in the sample depends on the correctness of your sampling and distributional assumptions.

EXERCISE 3A

In this exercise, you are expected to make a sensible choice of significance level for the hypothesis tests involved. Remember that the 5% level is conventional in scientific contexts.

1 Fourteen rats were timed as they ran through a maze. In one condition, the rats were hungry; in the other, they had just been fed.

(i) Use a paired *t*-test to test the hypothesis that the rats run the maze more quickly when they are hungry. The data below give the rats' times in each condition.

Rat	A	B	C	D	E	F	G	H
Fed time (seconds)	30	31	25	23	50	26	14	27
Hungry time (seconds)	29	18	14	27	37	34	15	22

Rat	I	J	K	L	M	N
Fed time (seconds)	31	39	38	39	44	30
Hungry time (seconds)	29	18	20	10	30	32

(ii) Do you think the assumptions for the paired *t*-test are justified here?

(iii) Half the rats were made to run the maze first when hungry and half ran it first when fed. Why did the experimenter do this?

2 Twelve voters, who form part of a newspaper's public opinion panel, are asked to rate the Prime Minister's performance on a scale from one to ten, before and after the party conference.

(i) Test, using a paired *t*-test, the hypothesis that the Prime Minister's rating has improved after the party conference. The panel's ratings are given in the table below.

Voter number	1	2	3	4	5	6
Rating before	4	7	8	3	4	5
Rating after	9	6	7	7	6	6

Voter number	7	8	9	10	11	12
Rating before	2	1	2	5	3	5
Rating after	3	2	6	8	5	5

(ii) Assuming that these voters are a random sample of the electorate, do you think that the assumptions of the test are justified?

3 Two voters are asked on eight separate occasions to rate the Prime Minister's performance, on a scale from one to ten.

(i) Test, using a paired *t*-test, the hypothesis that the voters' ratings of the Prime Minister do not differ, on average. The ratings are as follows.

Month	Rating of voter 1	Rating of voter 2
Feb	4	5
Mar	4	3
Apr	5	4
May	7	9
Jun	9	7
Jul	10	8
Sep	8	4
Oct	10	9

(ii) Explain clearly the difference between this question and the previous one. What is the random variable in each case? What is the population? What are the assumptions for this test to be valid? Are they different from the assumptions you needed to make in question 2?

4 Seventeen subjects were given a test of concentration: to go through a written passage crossing out every 'a' and ringing any 't'. They were timed on this task. Each was then given a placebo (a harmless pill containing no active ingredients) which they were told would enhance their concentration and hence their speed on the test; the test was then repeated (with a different written passage).

(i) Using a paired *t*-test, test the hypothesis that the placebo has the effect of improving subjects' performance. The times for each subject are given in the table below.

Subject	A	B	C	D	E	F	G	H	I
Original time (seconds)	54	27	55	51	53	56	77	62	18
Placebo time (seconds)	36	51	45	29	70	56	44	30	41

Subject	J	K	L	M	N	O	P	Q
Original time (seconds)	18	46	61	43	59	34	20	72
Placebo time (seconds)	47	41	21	26	72	16	25	47

(ii) Do you consider that the assumptions appropriate to the *t*-test are justified?

(iii) Criticise the experiment.

5 A new computerised job-matching system has been developed which finds suitably-skilled applicants to fit notified vacancies. It is hoped that this will reduce unemployment rates, and a trial of the system is conducted in seven areas.

(i) Using a paired *t*-test, test the hypothesis that the new system reduces the rate of unemployment. The unemployment rates in each area just before the introduction of the system and after one month of its operation are recorded in the table below.

Area	Rate before new system (%)	Rate after new system (%)
1	10.3	9.3
2	3.6	4.1
3	17.8	15.2
4	5.1	5.0
5	4.6	3.3
6	11.2	10.3
7	7.7	8.1

(ii) Do you think that the assumptions for the *t*-test would be justified here?

(iii) What do you think is wrong with this trial of the system?

6 Two timekeepers at an athletics track are being compared. They each time the nine sprints one afternoon.

(i) Using the paired *t*-test, test the hypothesis that the two timers are equivalent, on average. The times they record are listed below.

Race	1	2	3	4	5
Timer 1	9.65	10.01	9.62	21.90	20.70
Timer 2	9.66	9.99	9.44	22.00	20.82

Race	6	7	8	9
Timer 1	20.90	42.30	43.91	43.96
Timer 2	20.58	42.39	44.27	44.22

(ii) Are the assumptions appropriate to the *t*-test justified in this case?

The *t*-test with unpaired samples

The members of a maths class were asked one morning to check the time shown by their watches, then look away and, when they estimated that a minute had elapsed, to check their watches again to see how long had in fact elapsed.

The same procedure was followed with another class, from the same year group, that afternoon. The back-to-back stem-and-leaf diagram below shows the results.

Morning class		Afternoon class
(24 students)		(22 students)
	2	8
2	3	1
5 5 6	3	6
	4	4 4 0
7 8	4	9 9 9 6 6 5
1 1 3 3 3 3 4 4	5	2 2 1 1 1
5 5 5 7 7 7 7	5	8 7 7 5
1 4 4	6	3
	6	3 represents 63 seconds

This experiment gives another set of data with which you could test the hypothesis given at the start of the chapter. This time the experiment has an *unpaired design*: two separate groups of subjects are used in the two conditions.

EXPERIMENT

Try the experiment above: you will need reasonably large groups.

This section uses the data given in the stem-and-leaf diagram above to work through the process of hypothesis testing in this new context. If you have done the experiment above, you might find it helpful to follow the calculations with your data.

You cannot look here at the difference between a before-lunch and an after-lunch time: there is no way of associating particular before-lunch times with particular after-lunch times; but you can look at the difference between the mean before-lunch time and the mean after-lunch time. In fact, you can make the hypotheses:

H_0: There is no difference between the mean of people's estimates of one minute before and after lunch.

H_1: After lunch, the mean of people's estimates of one minute tends to be shorter than before lunch.

You can then use as your sample statistic the difference between the before-lunch sample mean and the after-lunch sample mean. As before, you now need to calculate the distribution of this sample statistic on the assumption that the null hypothesis is true.

The test statistic and its distribution

Assume that each before-lunch estimate is an independent random variable X_i $(i = 1, \ldots 24)$ with the normal distribution $N(\mu_b, \sigma_b^2)$ and each after-lunch estimate is an independent random variable Y_j $(j = 1, \ldots, 22)$ with normal distribution $N(\mu_a, \sigma_a^2)$. You are also making the assumption that each X is independent of each Y. This is a plausible assumption; it merely requires the independence of the two samples taken.

Recall that the mean of a sample of size n from a normal distribution $N(\mu, \sigma^2)$ has distribution:

$$N\left(\mu, \frac{\sigma^2}{n}\right).$$

In this case, therefore, the mean of the 24 before-lunch estimates has distribution:

$$\overline{X} \sim N\left(\mu_b, \frac{\sigma_b^2}{24}\right)$$

and the mean of the 22 after-lunch estimates has distribution:

$$\overline{Y} \sim N\left(\mu_a, \frac{\sigma_a^2}{22}\right).$$

Next you need a result that was revised in Chapter 2: If X has distribution $N(\mu_X, \sigma_X^2)$ and Y has distribution $N(\mu_Y, \sigma_Y^2)$ then $(X - Y)$ has distribution $N(\mu_X - \mu_Y, \sigma_X^2 + \sigma_Y^2)$.

Here, the distribution of the differences of the two sample means is therefore:

$$\overline{X} - \overline{Y} \sim N\left(\mu_b - \mu_a, \frac{\sigma_b^2}{24} + \frac{\sigma_a^2}{22}\right).$$

The null hypothesis then states that both means are equal, i.e. $\mu_b = \mu_a$ and so, if the null hypothesis is true:

$$\overline{X} - \overline{Y} \sim N\left(0, \frac{\sigma_b^2}{24} + \frac{\sigma_a^2}{22}\right).$$

Unfortunately, you do not know σ_b^2 or σ_a^2, so you will want, as you have in earlier work, to replace these unknown values with sample estimates. It might seem most natural to use separate estimates for the two unknown variances, but in fact it turns out that it is then hard to make any progress in calculating the distribution. This chapter, therefore, only deals with the case where you can assume that the variances in the two conditions are equal: here, this means that $\sigma_b^2 = \sigma_a^2 = \sigma^2$; that is, the before- and after-lunch estimates have the same variance. In this case:

$$\overline{X} - \overline{Y} \sim N\left(0, \sigma^2\left(\frac{1}{24} + \frac{1}{22}\right)\right).$$

So that:

$$\frac{\overline{X} - \overline{Y}}{\sqrt{\sigma^2\left(\frac{1}{24} + \frac{1}{22}\right)}} = \frac{\overline{X} - \overline{Y}}{\sigma\sqrt{\frac{1}{24} + \frac{1}{22}}} \sim N(0, 1). \qquad \text{①}$$

To estimate σ^2, you can use the pooled variance estimator from the two samples, using a method which will be discussed in Chapter 7.

$$S^2 = \frac{(24 - 1)S_b^2 + (22 - 1)S_a^2}{(24 + 22 - 2)}$$

where S_b^2 and S_a^2 are the usual unbiased sample estimators of the population variance given by:

$$S_b^2 = \frac{\sum_{i=1}^{24}(X_i - \overline{X})^2}{23} = \frac{\sum_{i=1}^{24} X_i^2 - 24\overline{X}^2}{23} \quad \text{and} \quad S_a^2 = \frac{\sum_{i=1}^{22}(Y_i - \overline{Y})^2}{21} = \frac{\sum_{i=1}^{22} Y_i^2 - 22\overline{Y}^2}{21}$$

The test statistic,

$$\frac{\overline{X} - \overline{Y}}{S\sqrt{\frac{1}{24} + \frac{1}{22}}}$$

which is obtained from ① by replacing the value of σ^2 with its estimator S^2, then has a t-distribution, with number of degrees of freedom equal to that in the pooled variance estimate: $(24 + 22 - 2) = 44$.

Carrying out the *t*-test for an unpaired sample

In the example:

$$\overline{x} = 51.542, \ s_b = 8.797; \ \overline{y} = 47.909, \ s_a = 8.574$$

so:

$$s = \sqrt{\frac{23 \times 8.797^2 + 21 \times 8.574^2}{44}} = 8.691$$

and the value of the test statistic is:

$$\frac{51.542 - 47.909}{8.691 \times \sqrt{\frac{1}{24} + \frac{1}{22}}} = 1.416$$

The critical region for a one-tailed test, in the case of 44 degrees of freedom, at the 5% significance level is $t > 1.680$ (this value does not appear in the tables, but can be obtained by interpolation for the values given for 30 and 50 degrees of freedom). Since $1.416 < 1.680$, the results lead you to accept the null hypothesis at this significance level: there is no good evidence that the before-lunch mean and the after-lunch mean of the population as a whole are different.

The process described is a *t-test for the difference of two means with unpaired samples*.

EXPERIMENTS

Carry out at least one of the experiments below, and test the hypothesis given using an unpaired sample *t*-test.

1 Use your reaction timer to decide whether males and females have the same mean reaction times, or whether older people have slower reactions than young people (you can choose the definition of 'older' to suit the samples you have available) or whether squash players have quicker reactions than non-players.

2 Are students studying A-level maths better at mental arithmetic than those taking other A-levels? You will need to devise a mental arithmetic test (Do you want to test speed or accuracy?) and administer it to a group of A-level maths students and a group of students taking other A-levels. *Do not be disappointed by the results! You can adapt this test to suit your prejudices: are A-level geography students better at naming capitals of foreign countries? Are A-level English students better at spelling?*

3 Two groups of subjects are each given lists of 25 words. Both groups must run down the list as quickly as possible. Those in the first group tick the words that are in capital letters. (You should make sure that about half of the words, placed randomly in the list, are in capital letters.) The second group ticks the words which rhyme with a target word which you give them. (Make sure that about half of the words, placed randomly in the list, do rhyme with this word.) You then ask the subjects each to write down as many words as they can remember from the list: do not tell the subjects in advance that they will have to do this. Test the hypothesis that the subjects who have looked for rhymes remember more of the words than those who looked for capital letters. Why would it be difficult to run this experiment with a paired design?

4 It would be useful to repeat whichever experiment, from those on pages 51–52, you carried out with a paired *t*-test. However, this time you should devise an unpaired design – that is, choose a separate set of subjects for each condition. This should help you to see the difference between the tests – and why the paired design is often preferable.

Assumptions for the unpaired *t*-test

The distributional assumptions needed for the unpaired *t*-test are quite severe.

1 The random variables measured in the two conditions must be independent.

2 They must be normally distributed.

3 They must have equal variances in the two different conditions.

Are these assumptions justified? The only information you have to help you decide is the two samples: the stem-and-leaf diagram for the data of the example is shown again below.

Morning class		Afternoon class	
(24 students)		(22 students)	
	2	8	
2	3	1	
5 5 6	3	6	
	4	4 4 0	
7 8	4	9 9 9 6 6 5	
1 1 3 3 3 3 4 4	5	2 2 1 1 1	
5 5 5 7 7 7 7	5	8 7 7 5	
1 4 4	6	3	
	6	3	represents 63 seconds

At first sight, the distributions here do not look much like samples from a normal distribution: they are rather obviously negatively skewed. Neither is it clear that they would have come from populations of the same variance.

You do not look further at these problems here, but if you are studying *Statistics 5*, you will be meeting a test for equality of variances; in Chapter 4 you will meet a test for a difference in location between two conditions which does not require the normality assumption.

 Do you think the assumptions made in the unpaired *t*-test are justified in the case of the experiment you carried out?

Comparison between paired and unpaired *t*-tests

The result we obtained above may seem a little surprising. If you compare the data from the paired experiment at the start of this chapter with that from the unpaired experiment you have just been analysing, you will see that the before- and after-lunch times in each case appear to have very similar distributions, though, in the paired case, with a considerably smaller sample, which would normally lead to a less decisive test.

	Data from paired experiment	**Data from unpaired experiment**
Before-lunch times	$\bar{x} = 51.250,\ s_b = 8.237,\ n = 12$	$\bar{x} = 51.542,\ s_b = 8.797,\ n = 24$
After-lunch times	$\bar{y} = 47.583,\ s_a = 9.258,\ n = 12$	$\bar{y} = 47.909,\ s_a = 8.574,\ n = 22$

Why then do you accept the null hypothesis in the paired case where the sample size is considerably smaller which, all other things being equal, would normally lead to a less decisive test?

You can see why the opposite appears to have happened if you look at the test statistics for the two cases.

Test statistic for paired experiment	**Test statistic for unpaired experiment**
$\dfrac{51.250 - 47.583}{6.946\sqrt{\frac{1}{12}}} = 1.829$	$\dfrac{51.542 - 47.909}{8.691\sqrt{\frac{1}{24} + \frac{1}{22}}} = 1.416$

The test statistics for the paired and unpaired calculations have very similar numerators, but the standard error in the denominator is considerably larger in the unpaired calculation, despite the larger sample size in that case.

The crucial point is that there is, for all sorts of reasons, considerable variation amongst people in their reaction times and lunch is only one, relatively small, effect amongst many. Some people will tend to make short estimates in both conditions and some long estimates in both conditions, though in both cases the effect of lunch may be the same.

The paired design enables us to take this into account in a way which the unpaired design cannot because of the way the standard error is estimated.

- The standard error for the paired experiment estimates how much variation to expect in the average difference between *a particular person's* before- and after-lunch times. It is calculated from the standard deviation of these differences in the sample and is therefore based only on the variation in the effect of lunch on different people, not on the values s_b and s_a which describe the very considerable variation in reaction times between people for all the other reasons.

- The standard error for the unpaired experiment estimates how much variation to expect in the average difference between *a random before-lunch time and a random after-lunch time*. It is calculated from the values s_b and s_a which take into account the very considerable variation between people for all sorts of reasons, including the relatively small variation associated with lunch.

Using paired and unpaired *t*-tests

It is a characteristic of research by social scientists that they are looking for a small average difference between the values of a particular variable in two different conditions, but that subjects show very substantial variation in the values of this variable within both conditions. In these situations, an unpaired *t*-test is not usually very helpful, as it will require a very large sample size to discriminate between the null hypothesis of no difference between the means in the two conditions and the true situation where there is a small difference. If you study *Statistics 5* you will see that this is saying that the unpaired test is not very powerful.

Considerable ingenuity is therefore employed in attempting to match subjects so that a paired test can be used to eliminate some of the variation between them and the small difference between the two experimental conditions is not swamped.

In the paired experiment you used the same subject in each of two conditions, but this is not necessary. In fact, having taken part in one experimental condition sometimes makes it impossible to take part in the second.

For example, if you wish to test the effect on children's intelligence of an upbringing in families from two different social classes you could not use the same child and bring it up twice, nor would an unpaired *t*-test be suitable in this case: the variation in intelligence caused by other factors would swamp the effect you are looking for.

One possibility is to find pairs of identical twins who are being adopted at birth and are assigned to adoptive parents of different social classes: these constitute matched pairs of subjects and you could use a *t*-test on the differences between the intelligences of the twins from the two types of family. Notice that here the matching is perfect in the sense that both children have identical genetic endowments: the belief implicit in this experiment is that heredity is a major cause of variation in intelligence and this effect will be cancelled out by the matching process. Of course, there will be many differences between the adoptive families other than class, and it is possible that the variations in intelligence induced by these differences in upbringing will still swamp the effect being examined. Ideally, you would want to find identical twins being assigned to families differing only in their social class, but it is unlikely that you would find enough, if any, examples of this to conduct the test!

In this exercise, you are expected to make a sensible choice of significance level for the hypothesis tests involved. Remember that the 5% level is conventional in scientific contexts.

1 A species of finch has subspecies on two different Galapagos Islands.

The weights of a sample of finches from each island are listed below.

Weights from Daphne Major (grams)

64	67	64	61	68	61	61	67	70	62	66	63

Weights from Daphne Minor (grams)

63	61	65	65	61	63	63

(i) Is there evidence that the finches on Daphne Major are heavier on average than those on Daphne Minor?

(ii) What assumptions do you need to make? Are they reasonable?

2 Two different varieties of sprout being grown around the country, on a variety of different plots with different soil types and weather conditions, are monitored by a crop-testing station.

(i) Test, using an unpaired *t*-test, the hypothesis that the two varieties have the same average yield. The yields of sprouts per square metre in kilograms are as follows.

Yields of 'Old Cobby' (kg/m^2)

18	15	14	21	21	19	9	20	26	27
21	12	15	14	27	22	23	18	10	19

Yields of 'Early Yellow' (kg/m^2)

12	20	18	14	15	13	17	15	20	26
17	14	8	29	24	16	20	20	18	23
20	23	23	18	9					

(ii) Plot the data on a back-to-back stem-and-leaf diagram. State and comment on the assumptions you are making in carrying out the test. Do you think the assumptions are justified?

(iii) Devise an experiment to test this hypothesis which would use a paired *t*-test. Would this be an improvement, do you think?

3 Two groups of subjects are asked to volunteer for a psychology experiment. One group is told that they will be paid £1 for participating, the other that they will be paid £20. The experiment consists of a rather dull task which must be repeated for one hour: subjects are then asked to rate how interesting the task was, on a scale from 1 to 10; 10 being the most interesting.

(i) Test, using an unpaired *t*-test, the hypothesis that the task was found more interesting by those who were paid less.

The ratings of the two groups were:

Paid	4	2	6	3	5	1	6	4
£1	3	5	5	3	2	2	6	
	1	7	2	4	5	4	5	
Paid	3	5	2	1	3	4	5	
£20	2	1	1	3	4	4	4	
	5	2	3	2	1	3		

(ii) State and comment on the assumptions you are making in order to carry out this test.

(iii) Could you devise a paired design for this experiment?

4 Some personality theorists classify people into 'introverts', or quiet personalities, and 'extroverts', or outgoing personalities. In an attempt to find physical correlates of these personality types, the heights of a group of 17 extroverts and a group of 17 introverts were measured.

(i) Show that an unpaired *t*-test with the data below accepts, at the 5% level, the hypothesis that extroverts are taller on average than introverts, stating carefully the assumptions you are making.

Heights of extroverts (cm)

179	174	166	170	170	168
173	170	173	173	168	174
163	156	157	162	169	

Heights of introverts (cm)

159	172	167	168	163	155
164	163	166	164	175	155
165	168	164	166	162	

(ii) The scientist who made these measurements concluded that taller people tend to be more confident and hence extroverted. Is this conclusion justified?

5 A college careers officer investigates the income at age 24 of a group of students who left school at 16, and a group who stayed on to take A-levels.

The results he finds (measuring incomes in £ per week) are summarised by this table.

	16-year-old leavers	A-levels leavers
Mean	156	164
Sample estimate of variance	673	593
Sample size	37	28

(i) Show, using an unpaired t-test, that the hypothesis that those staying on at school have higher incomes at age 24 is rejected, on the evidence of this sample. What assumptions are you making for an unpaired t-test to be appropriate? How plausible are they?

(ii) What other difference between the two groups inevitably exists that might explain this unexpected result? How could you design an experiment to eliminate this effect?

6 My sister is worried that the office in which she works suffers from 'sick building' syndrome. She has noticed that those who work on her floor, which is air-conditioned, seem to suffer from colds more often than those on the non-air-conditioned floor below. She uses the sick-leave records to discover the number of days of absence with colds in the last year.

(i) Use an unpaired t-test to test the hypothesis that staff on the air-conditioned floor have more absences with colds, using the following data.

Absence of staff:

on her floor	6	5	11	2	7	13	8
	10	6	1	0	8	3	2
on the floor below	2	0	5	5	7	1	
	0	3	8	2	4		

(ii) What assumptions are necessary for the t-test to be appropriate? Which of them are not justified here?

Testing for a non-zero value of the difference of two means

You have now used two different versions of the t-test to examine the null hypothesis that two different conditions produce the same mean value of some random variable. As in *Statistics 3*, the method can also be used in a more gener¹ way to test null hypotheses which suggest that the mean of a random variable, differs by a given amount in the two conditions.

Paired test

Null hypothesis: for some given value of δ

H_0: The mean difference between the values of X in condition 1 and condition 2 is δ.

Sample: A set of pairs, one in each condition, of observations of X.

Let U be the random variable obtained by subtracting the corresponding value of X in the two conditions and n the number of pairs of observations.

You use these values to calculate the sample mean $\overline{U}$ and the unbiased sample estimator S^2 of the population variance of U. Then:

$$\frac{\overline{U} - \delta}{\frac{S}{\sqrt{n}}} \sim t_{n-1}$$

provided that U is distributed normally in the population and that the null hypothesis is true.

EXAMPLE 3.1

In the population as a whole, the mean difference in heights of 17-year-old boys and girls is 4.3 cm. A scientist suspects that this difference would not be the same if she looked at boys and girls who were (non-identical) twins. Her hypotheses are:

H_0: The mean difference between 17-year-old twin boy and girl heights is 4.3 cm.

H_1: The mean difference between 17-year-old twin boy and girl heights is different from 4.3 cm.

She assumes that the difference between twin boy and girl heights is modelled by a normal random variable.

The data she has collected are shown in this table.

Should she accept the null hypothesis?

Family	A	B	C	D	E	F	G	H
Boy's height (cm)	176	157	170	166	162	175	157	171
Girl's height (cm)	156	162	163	168	159	174	170	165
Difference (cm)	20	−5	7	−2	3	1	−13	6

SOLUTION

The average difference is $\overline{u} = 2.125$ and the sample estimate of the population standard deviation is $s = 9.687$

The test statistic is therefore

$$\frac{2.125 - 4.3}{\frac{9.687}{\sqrt{8}}} = -0.635$$

and there are (size of sample -1) $= (8 - 1) = 7$ degrees of freedom.

The critical region for a two-tailed test with seven degrees of freedom at the 5% significance level is $t > 2.365$ or $t < -2.365$, so that in this case, since $-2.365 < -0.635 < 2.365$, she accepts the null hypothesis that the mean difference between heights of twin boys and girls is 4.3 cm.

Unpaired test

Hypothesis: for some given value of δ

H_0: The difference between the mean values of X in condition 1 and condition 2 is δ.

Sample: Two sets of observations of X, one set in each condition.

Let X_1 and X_2 be the random variables in the two conditions and n_1 and n_2 be the number of observations under each condition.

Use these values to calculate the sample means $\overline{X}_1$ and $\overline{X}_2$ and the unbiased pooled-sample estimator S^2 of the population variance.

Then:

$$\frac{(\overline{X}_1 - \overline{X}_2) - \delta}{S\sqrt{\dfrac{1}{n_1} + \dfrac{1}{n_2}}} \sim t_{n_1 + n_2 - 2}$$

provided that the random variable X is distributed normally in the population, with the same variance in each condition, and that the null hypothesis is true.

EXAMPLE 3.2

The manufacturers of a dieting compound 'Slimplan', claim that the use of their product as part of a calorie-counting diet leads to an average extra weight loss of at least five pounds in a fortnight. An experiment has been carried out by a consumers' group which doubts this claim.

The hypotheses are:

H_0: The mean extra weight loss in a fortnight from adding 'Slimplan' to a calorie counting diet is five pounds.

H_1: The mean extra weight loss in a fortnight from adding 'Slimplan' to a calorie counting diet is less than five pounds.

The assumptions are that the weight loss in a fortnight from a calorie counting diet, with or without 'Slimplan', is a normally distributed random variable and that the addition of 'Slimplan' to the diet does not affect the variance of this random variable.

Thirty-six dieters used 'Slimplan' with their diets; their weight losses x_i $(i = 1, \ldots, 36)$ in pounds are summarised by the figures:

$$\sum_{i=1}^{36} x_i = 409.32 \qquad \sum_{i=1}^{36} x_i^2 = 6102.39.$$

Sixty-two dieters followed the same calorie-counting procedure, but did not use 'Slimplan'; their weight losses y_j $(j = 1, \ldots, 62)$ in pounds are summarised by the figures:

$$\sum_{j=1}^{62} y_j = 571.64 \qquad \sum_{j=1}^{62} y_j^2 = 5618.40.$$

SOLUTION

These data give: $\bar{x} = 11.37$, $s_x = 6.433$, $\bar{y} = 9.22$ and $s_y = 2.38$

so that

$$s = \sqrt{\frac{35 \times 6.433^2 + 61 \times 2.388^2}{36 + 62 - 2}} = 4.326$$

The test statistic is:

$$\frac{(11.37 - 9.22) - 5}{4.326 \times \sqrt{\dfrac{1}{36} + \dfrac{1}{62}}} = -3.144$$

and there are (size of sample 1 + size of sample 2 − 2) = (36 + 62 − 2) = 96 degrees of freedom.

The critical region for a one-tailed test with 96 degrees of freedom at the 5% significance level is $t < -1.661$ and so, since $-3.144 < -1.661$ the null hypothesis is rejected in favour of the alternative that the average extra weight loss is not as great as five pounds.

EXERCISE 3C

In this exercise, you are expected to make a sensible choice of significance level for the hypothesis tests involved. Remember that the 5% level is conventional in scientific contexts.

1 Amongst all praying mantises, females are on average 7 cm longer than males. A new variety of mantis has been bred, the insects of which are supposed to be more nearly equal in size.

Test the hypothesis that the difference between male and female average lengths is less than 7 centimetres, using the lengths in centimetres of

the sample of twelve males and twelve females shown below. State clearly the assumptions you are making in your test.

Males	18.4	15.1	11.9	16.3	8.7	13.0
	10.1	20.2	14.2	6.2	16.9	8.8
Females	28.2	24.2	12.4	23.4	13.2	21.4
	10.3	23.5	16.8	11.9	15.5	13.2

2 In fact, the data given in question 1 are paired: each male mantis is paired with its mate in the following way.

Pair	1	2	3	4	5	6
Male	18.4	15.1	11.9	16.3	8.7	13.0
Female	28.2	24.2	12.4	23.4	13.2	21.4

Pair	7	8	9	10	11	12
Male	10.1	20.2	14.2	6.2	16.9	8.8
Female	10.3	23.5	16.8	11.9	15.5	13.2

(i) Test the hypothesis that male mantises are on average less than 7 centimetres smaller than their mates.

(ii) Explain clearly what assumptions you make in this case, and how these assumptions differ from those you made in question 1.

(iii) Why are the results you obtain different in this case from those you found in question 1?

3 It is known from many studies that the best current post-operative treatment reduces stays in hospital after major operations, compared with untreated patients, by an average of 6.2 days. A new treatment is proposed, with the hypothesis that this new treatment will reduce stays in hospital by more than 6.2 days on average, and a trial is conducted on two groups of patients who have just undergone major operations. The results are shown below.

Days stayed in hospital by untreated patients	35	33	27	27	25
	31	27	36	46	
	32	27	16	28	
Days stayed in hospital by patients given new treatment	18	18	23	24	
	22	14	28	29	
	24	16	18		

Test the hypothesis given, stating the assumptions you are making clearly.

4 In 1979, the world price of crude oil rose by 60%. My macroeconomic model suggests that such a rise should result in an increase in the inflation rate of 2% between 1979 and 1980. The data I have available on inflation rates in these years are shown in the table below.

Country	Inflation rate 1979 (%)	Inflation rate 1980 (%)
Australia	9.1	10.2
Italy	14.8	21.2
Japan	3.7	7.7
Spain	15.6	15.6
UK	13.4	18.0
USA	11.3	13.5
Germany	4.1	5.5

(i) Test the hypothesis that the inflation rate rose by 2% between 1979 and 1980.

(ii) State carefully the assumptions you are making. What is a random sample in this context?

(iii) Do you think this is an adequate test of my model? Explain.

5 Given a list of 30 three-letter 'words' to learn, an adult can recall an average of 10.3 more 'words' if they are genuine English words than if they are random 'nonsense' combinations of three letters. It is thought that in dyslexic adults, however, there is a larger difference in recall ability between real and nonsense 'words'. An experiment to test this is carried out. Ten dyslexic adults are given lists of real and nonsense 'words', and their recall tested. The results are listed in the table below.

Initials	TP	CS	AD	JM	MC
Real words recalled	17	19	24	19	16
Nonsense words recalled	4	5	6	6	12

Initials	AB	RS	PT	LL	MR
Real words recalled	20	25	29	22	15
Nonsense words recalled	3	15	18	10	1

Test the hypothesis that these dyslexic adults recall more than 10.3 fewer nonsense 'words' on average than real words, stating the assumptions you are making.

6 When learning real words, dyslexic adults recall on average 5.1 fewer of a list of 30 words than adults who are non-dyslexic. In an experiment to decide whether this is true also of nonsense words, a group of dyslexic and a group of non-dyslexic adults are given a list of nonsense 'words' to learn. When their recall is tested the numbers of nonsense words recalled were as follows.

	by dyslexic adults	by non-dyslexic adults
mean	5.2	12.8
sample estimate of variance	36.7	45.1
sample size	28	41

(i) Test the hypothesis that dyslexic adults recall on average 5.1 fewer nonsense 'words' than those who are non-dyslexic.

(ii) State carefully the assumptions necessary, and explain how these differ from the assumptions of question 5.

(iii) State clearly the relation between this hypothesis and the hypothesis of question 5. Are they related at all? If so, how?

1 Given an experiment which produces pairs of values (X, Y) of two random variables, where $(X - Y)$ is normally distributed, to conduct a *paired t-test* with null hypothesis

$$H_0: E(X) = E(Y) + \delta$$

using as data n pairs (x_i, y_i) whose differences $(x_i - y_i)$ provide an independent random sample of values of $(X - Y)$:

- calculate the difference $(x_i - y_i)$ for each data pair
- find the mean $\overline{d}$ and unbiased variance estimate s^2 from the differences
- calculate the test statistic

$$\frac{\overline{d} - \delta}{\frac{s}{\sqrt{n}}}$$

- compare with the appropriate critical value of the *t*-distribution with $(n - 1)$ degrees of freedom.

2 Given an experiment which produces sets of values of two random variables X and Y, each of which is normally distributed with the same variance, to conduct an *unpaired t-test* with null hypothesis:

$$H_0: E(X) = E(Y) + \delta$$

using as data m values x_i and n values y_i which constitute an independent random sample of X and Y, respectively:

- find the means $\overline{x}, \overline{y}$ of the data sets
- calculate the unbiased estimate of the common variance s^2 from the pooled samples
- calculate the test statistic

$$\frac{\overline{x} - \overline{y} - \delta}{s\sqrt{\dfrac{1}{m} + \dfrac{1}{n}}}$$

- compare with the appropriate critical value of the *t*-distribution with $(m + n - 2)$ degrees of freedom.

Confidence intervals

And, confident we have the better cause, why should we fear the trial?

Philip Massinger

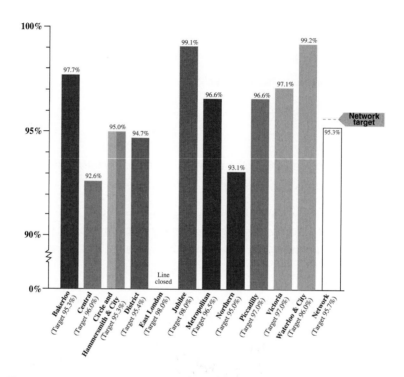

Figure 4.1 How do you think these London Underground lines set their targets?

A transport company is tendering for a contract to carry loads over a particular route on a regular basis. In order to estimate its costs accurately the company wants to know a value which it can be fairly sure the mean journey time will not exceed.

Lorries have made trial runs over the route seven times, recording the following journey times in minutes:

$$168, \ 192, \ 245, \ 160, \ 203, \ 184, \ 211.$$

The average of this sample of journey times is $\bar{t} = 194.7$ minutes, so the company might use this as an estimate of the mean journey time.

A single value suggested for the mean in this way is called a *point estimate* of the parameter but a more useful process is often to give an *interval estimate*, which is a range of suggested values in which the parameter might lie. In *Statistics 3* you saw how to calculate a symmetrical confidence interval for the mean of a distribution. This gave both upper and lower limits to the values that the mean might reasonably be expected to have. Here, you will go through a very similar construction but this time using the data to put only an upper limit on the values that the mean might reasonably be expected to have.

This is a more appropriate procedure for this problem, because the company is interested in finding a value which it can be fairly sure the mean journey time will not exceed: it is not concerned with a lower bound.

The procedure for constructing a one-sided confidence interval is very similar to that for a two-sided interval.

- From the data calculate the sample mean $\bar{t}$ and the unbiased sample estimate s^2 of the population variance. Here, the values are:

$$\bar{t} = 194.7, s = 28.6.$$

- The estimated standard error of the sample means is $\dfrac{s}{\sqrt{n}}$ where n is the sample size. Here, this is:

$$\frac{s}{\sqrt{7}} = 10.8.$$

- The *upper confidence bound* for the population mean is then the appropriate number of standard errors above the sample mean: that is, $\bar{t} + \tau \dfrac{s}{\sqrt{n}}$, where τ is the relevant critical value of the t-distribution with $n - 1$ degrees of freedom.

Here, you will find a 95% confidence interval, so the one-tailed critical value for the 5% level of significance is appropriate. With six degrees of freedom, this value is 1.943, so the upper confidence bound is:

$$\bar{t} + 1.943\frac{s}{\sqrt{7}} = 215.7.$$

An equivalent way of expressing this result is to say that the range $(-\infty, 215.7)$ is a 95% one-sided confidence interval for the population mean.

JUSTIFICATION

If the random variable T representing the journey time is normally distributed with mean μ and variance σ^2, then you know that the variable $\bar{T}$ which represents the mean of a random sample of seven journey times has the distribution:

$$\bar{T} \sim N\left(\mu, \frac{\sigma^2}{7}\right).$$

Then:

$$\frac{\bar{T} - \mu}{\frac{\sigma}{\sqrt{7}}} \sim N(0, 1)$$

and it follows that:

$$\frac{\bar{T} - \mu}{\frac{S}{\sqrt{7}}} \sim t_6$$

where S^2 is the unbiased estimator of σ^2, calculated from the seven times in the sample.

The fact that 1.943 is the one-tailed 5% critical value of the t-distribution with six degrees of freedom means that:

$$P\left(-1.943 < \frac{\overline{T} - \mu}{\dfrac{S}{\sqrt{7}}}\right) = 0.95. \qquad \qquad ①$$

Note that the probability here refers to possible outcomes of the whole process of taking a random sample and calculating the value of $\dfrac{\overline{T} - \mu}{\dfrac{S}{\sqrt{7}}}$ from it, when the actual values of μ and σ remain fixed.

One interpretation of ① is therefore that over many uses of this procedure, the value calculated for $\dfrac{\overline{T} - \mu}{\dfrac{S}{\sqrt{7}}}$ will exceed -1.943 in 95% of the samples taken.

Rearranging this gives the result:

$$P\left(\mu < \overline{T} + 1.943\,\frac{S}{\sqrt{7}}\right) = 0.95$$

which justifies the use of the value $\bar{t} + 1.943\,\dfrac{s}{\sqrt{7}}$ as an upper confidence bound for μ.

Interpretation of the confidence bound

The result at the end of the previous section says that, over many uses of this procedure, the value of $\overline{T} + 1.943\,\dfrac{S}{\sqrt{7}}$ obtained from a sample will exceed the true value of μ in about 95% of samples. By taking this value as an upper limit for an interval estimate of the true mean journey time, you are hoping that your sample will be one of these 95%.

It is important to realise that this result does not say that the true mean lies within the confidence interval with probability 0.95: still less does it justify claims such as '95% of the lorries make the journey in less than the upper confidence bound'. The correct interpretation is that if confidence intervals are constructed in this way, then they will, over many uses of the procedure, include the true mean 95% of the time.

How to choose the confidence level

In the example, you found the 95% confidence interval $(-\infty, 215.7)$. With the same data, the 99.5% confidence interval is $(-\infty, 234.8)$ and the 50% confidence interval is $(-\infty, 194.7)$. Notice that the higher the confidence level:

- the less likely it is that the interval fails to include the true mean
- the less restrictive the upper bound is.

The only way to get a tighter upper bound without reducing the chance of including the true mean is to collect a larger sample.

Therefore, in practice, the decision as to which confidence level to choose will depend on balancing the following three factors.

- The cost of basing the company's scheduling on a range of suggested mean journey times which does not include the true mean: if this cost is large, the company will want a higher confidence level or a larger sample size.

- The cost in wasted time of a large uncertainty in the mean journey time: if this cost is high the company will want a lower confidence level or a larger sample size.

- The cost of increasing the sample size.

Lower confidence bounds

A one-sided confidence interval can also, in a very similar way, be given as a lower $(100 - a)\%$ confidence bound for the population mean μ of a random variable X, based on a sample of size n. This is given by:

$$\bar{x} - \tau \frac{s}{\sqrt{n}} < \mu$$

where $\bar{x}$ is the sample mean, s^2 is the unbiased sample estimate of the population variance, and τ is the $a\%$ critical value for the t-distribution with $(n - 1)$ degrees of freedom.

Using the data from the example, a lower 95% confidence bound for the lorries' mean journey time is:

$$\bar{t} - 1.943 \frac{s}{\sqrt{7}} = 173.7.$$

Asymmetrical confidence intervals

The transport company in the example you have been considering might want to be fairly sure that the upper bound of possible mean times which it uses in its scheduling does include the true mean, because of the cost to the company of being unable to meet its delivery commitments. Thus the upper 99.5% confidence limit of 234.8 might be appropriate. It may, on the other hand, be less concerned about over-estimating the true mean time, because of the relatively lighter cost of slackness in the schedule. Thus a lower 95% confidence limit of 173.7 might be appropriate.

In this sort of situation, where the relative costs of over- and under-estimating the population mean are different, it may be appropriate to use different confidence levels for calculating upper and lower confidence bounds. In this case, you say that the region between these lower and upper bounds constitutes an *asymmetrical confidence interval* for the mean.

If the upper bound is at the $(100 - a)\%$ confidence level and the lower is at the $(100 - b)\%$ confidence level, then the upper bound is below the true mean with probability $a\%$ and the lower bound is above the true mean with probability $b\%$. Therefore the interval fails to include the true mean with probability $(a + b)\%$ and so the confidence level of the asymmetrical interval is said to be $(100 - (a + b))\%$.

In the example, the range $173.7 < \mu < 234.8$ is an asymmetrical confidence interval for the mean journey time with confidence level $100 - ((100 - 95) + (100 - 99.5)) = 94.5\%$.

EXAMPLE 4.1

A dairy company produces crème fraîche in tubs which are supposed to weigh 250 g. The weights of eleven tubs picked at random are summarised by the figures $\overline{w} = 248.26$ g, $s = 12.29$ g. Verify that the range (240 g, 260 g) is a 97% confidence interval for the true mean weight of crème fraîche in the tubs, stating any assumptions that underlie this statement.

SOLUTION

Using the formulae for the upper and lower confidence bounds, you must have:

$$\overline{w} - \tau_a \frac{s}{\sqrt{n}} = 248.26 - 3.706\tau_a = 240$$

$$\overline{w} + \tau_b \frac{s}{\sqrt{n}} = 248.26 - 3.706\tau_b = 260$$

where τ_a and τ_b are the critical values appropriate to the lower and upper bounds, respectively. Solving the equations gives:

$$\tau_a = 2.23$$

$$\tau_b = 3.17$$

From the t-tables, you can see that 3.17 is the one-tailed critical value for the 0.5% level, and 2.23 is the critical value for the 2.5% level, with ten degrees of freedom. The confidence level for the asymmetrical interval is therefore:

$$(100 - 2.5 - 0.5)\% = 97\%$$

as required.

The assumptions made are that an independent random sample was taken and that the underlying distribution of weights is normal.

1 The amounts of a particular impurity in eleven samples of a drug, measured in micrograms per gram of the drug, are listed below.

24 17 31 22 17 28 26 20 19 28 23

(i) Find a one-sided 97.5% confidence interval giving an upper limit for the average amount of impurity in the drug.

(ii) What are you assuming about:

(a) the distribution of amounts of the impurity in the drug

(b) the process of sampling?

2 Oil is being extracted from olives, and the amount that can be extracted from each olive is normally distributed.

(i) Explain the method for constructing confidence intervals which has the following properties.

- The true mean amount of oil extracted will lie above the upper limit of the interval in 5% of samples of size 10.

- The true mean amount of oil extracted will lie below the lower limit of the interval in 1% of samples of size 10.

(ii) Find the confidence interval in the case where a sample of ten olives produces the following amounts of oil (in millilitres).

| 0.443 | 0.412 | 0.505 | 0.487 | 0.444 |
| 0.401 | 0.387 | 0.461 | 0.453 | 0.428 |

3 Find a 99% one-sided confidence interval which provides a lower limit for the mean mass of adult rattlesnakes, given that a random sample of 17 adult rattlesnakes have masses (in grams) as listed below.

1229	1682	1452	1070	1992	1620	2103
1528	1701	1092	1203	2207	1852	1629
1603	1334	1835

4 Show that (6 m, 11 m) is a 98.5% confidence interval for albatross wingspans, given that six albatrosses have wingspans summarised by

$$\sum x = 49.65, \ \sum x^2 = 424.5.$$

Hypothesis tests and confidence intervals

There is a very close relationship between hypothesis tests and confidence intervals, which should be clearly understood.

A *hypothesis test* suggests a value for an unknown population parameter (the null hypothesis), and then accepts this value if a test statistic lies in a particular range (that is, it lies outside the critical region). However, the critical region depends on the hypothesised population parameter, so you can reverse this process. Thus for a given value of the test statistic, you can determine the range of values for the population parameter which would be accepted by the test if they were offered as null hypotheses. This is called the *confidence interval* for the population parameter.

For instance, in the case where you take an independent random sample of size n from a normal distribution to test the hypotheses:

H_0: Population mean $= \mu$

H_1: **(a)** Population mean $\neq \mu$

or **(b)** population mean $> \mu$

or **(c)** population mean $< \mu$.

The test statistic is:

$$\frac{\overline{x} - \mu}{\frac{s}{\sqrt{n}}}$$

and you accept the null hypothesis at the $p\%$ significance level if

(a) $-\tau_2 < \dfrac{\overline{x} - \mu}{\frac{s}{\sqrt{n}}} < \tau_2$ or **(b)** $\dfrac{\overline{x} - \mu}{\frac{s}{\sqrt{n}}} < \tau_1$ or **(c)** $\dfrac{\overline{x} - \mu}{\frac{s}{\sqrt{n}}} > -\tau_1$

where τ_1, τ_2 are the one- and two-tailed critical values respectively for the t-distribution with $n - 1$ degrees of freedom at the $a\%$ level.

Alternatively, for a given value of $\overline{x}$ you can view these inequalities as constraining the range of values of μ which would be accepted by the test if they were offered as null hypotheses, and rearranging them gives the $(100 - a)\%$ confidence intervals.

(a) $\overline{x} - \tau_2 \dfrac{s}{\sqrt{n}} < \mu < \overline{x} + \tau_2 \dfrac{s}{\sqrt{n}}$ or **(b)** $\overline{x} - \tau_1 \dfrac{s}{\sqrt{n}} < \mu$ or **(c)** $\overline{x} + \tau_1 \dfrac{s}{\sqrt{n}} > \mu$

You will see this relationship exploited in many different ways as you work through *Statistics 4* and *Statistics 5*.

Note

As with the point estimators you will look at in Chapter 7, the lower and upper bounds of a confidence interval are both random variables: the values they take depend on the sample that happens to be selected. This means that these bounds have sampling distributions with properties that will be of interest to statisticians, though they are not considered further here.

Confidence intervals for the difference of two means

The paired condition

In an experiment on group behaviour, twelve subjects were each asked to hold one arm out horizontally while supporting a 2 kg weight, under two conditions:

- while together in a group

- while alone with the experimenter.

The times, in seconds, for which they were able to support the weight under the two conditions were recorded as follows.

Subject	A	B	C	D	E	F	G	H	I	J	K	L
'Group' time	61	71	72	53	71	43	85	72	82	54	70	73
'Alone' time	43	72	81	35	56	39	63	66	38	60	74	52
Difference	18	−1	−9	18	15	4	22	6	44	−6	−4	21

The mean difference in the sample is $\bar{d} = 10.67$ and if you assume that the differences are normally distributed, then an unbiased estimate from the sample of the variance of this distribution is $s^2 = 232.24$.

The statistic

$$\frac{\bar{D} - \delta}{\dfrac{S}{\sqrt{12}}}$$

then has a t-distribution with eleven degrees of freedom, where δ is the true mean difference between 'group' and 'alone' times in the population as a whole.

The two-tailed critical value for the t-distribution with eleven degrees of freedom at the 10% level is 1.796. Thus:

$$P\left(-1.796 < \frac{\bar{D} - \delta}{\dfrac{S}{\sqrt{12}}} < 1.796\right) = 0.9 \text{ or}$$

$$P\left(\bar{D} - 1.796 \, \frac{S}{\sqrt{12}} < \delta < \bar{D} + 1.796 \, \frac{S}{\sqrt{12}}\right) = 0.9.$$

A 90% confidence interval for the mean difference between group and alone times is therefore:

$$\bar{d} - 1.796 \, \frac{S}{\sqrt{12}} < \delta < \bar{d} + 1.796 \, \frac{S}{\sqrt{12}}.$$

Remember what this means. If you take samples and calculate intervals from them in this way, the true difference between the 'group' and the 'alone' mean times will lie in this interval in 90% of the samples, over many uses of this procedure. Inserting the values from the data, the confidence interval is

$$2.77 < \delta < 18.57.$$

In general, suppose that pairs of values of a random variable, X, one in each of two conditions, are measured on a sample of size n. Let:

- D be the random variable obtained by subtracting the values of X in the two conditions and suppose that D is distributed normally in the population

- s^2 be the unbiased sample estimate of the population variance of D

- τ_a be the two-sided $a\%$ critical value for the t-distribution with $(n-1)$ degrees of freedom.

Then a $(100 - a)\%$ confidence interval for δ, the difference in mean X values between the two conditions, is given by:

$$\bar{u} - \tau_a \frac{s}{\sqrt{n}} < \delta < \bar{u} + \tau_a \frac{s}{\sqrt{n}}.$$

You should note the similarity between this confidence interval and the hypothesis test discussed in Chapter 3, on pages 49–51.

The unpaired condition

Two runners are being considered for a place in a team. They have each recently competed in several races, though not against each other. Their times (in seconds) were as shown in the table below.

Runner 1	47.2	51.8	48.1	47.9	49.0	48.2	48.1
Runner 2	49.5	47.4	48.3	49.1	47.6		

You can model the first and second runners' times with variables T_1 and T_2 with distributions $N(\mu_1, \sigma^2)$ and $N(\mu_2, \sigma^2)$, respectively. You are describing their running times as normally distributed with different means and a common variance. The different means reflect differences in the runners' underlying ability; the random variability comes from factors such as the influence of other runners and weather conditions for which the effects in the different races are independent.

Because you are interested in the difference in the runners' underlying abilities, you are looking for a confidence interval for the difference between μ_1 and μ_2.

The sample means of the runners' times have distributions:

$$\bar{T}_1 \sim N\left(\mu_1, \frac{1}{7}\sigma^2\right) \quad \text{and} \quad \bar{T}_2 \sim N\left(\mu_2, \frac{1}{5}\sigma^2\right)$$

so that the distribution of their difference is:

$$(\bar{T}_1 - \bar{T}_2) \sim N\left(\mu_1 - \mu_2, \sigma^2\left(\frac{1}{7} + \frac{1}{5}\right)\right).$$

The standardised variable

$$\frac{\left(\overline{T}_1 - \overline{T}_2\right) - (\mu_1 - \mu_2)}{\sigma\sqrt{\left(\dfrac{1}{7} + \dfrac{1}{5}\right)}}$$

then has an N(0, 1) distribution.

If you replace σ^2 with its unbiased sample estimator, $S^2 = \dfrac{(7-1)S_1^2 + (5-1)S_2^2}{(7+5-2)}$

where S_1^2 and S_2^2 are the unbiased sample estimators of the variance from the two separate samples, then, finally:

$$D = \frac{\left(\overline{T}_1 - \overline{T}_2\right) - (\mu_1 - \mu_2)}{S\sqrt{\left(\dfrac{1}{7} + \dfrac{1}{5}\right)}}$$

has distribution $t_{7+5-2} = t_{10}$.

The critical value for the t-distribution with ten degrees of freedom at the 5% significance level is 2.228, so that D lies between -2.228 and $+2.228$ in 95% of samples. That is, a 95% confidence interval for $(\mu_1 - \mu_2)$ is defined by:

$$-2.228 < \frac{\left(\bar{t}_1 - \bar{t}_2\right) - (\mu_1 - \mu_2)}{s\sqrt{\left(\dfrac{1}{7} + \dfrac{1}{5}\right)}} < +2.228.$$

This can be rearranged as:

$$\left(\bar{t}_1 - \bar{t}_2\right) - 2.228s\sqrt{\left(\dfrac{1}{7} + \dfrac{1}{5}\right)} < (\mu_1 - \mu_2) < \left(\bar{t}_1 - \bar{t}_2\right) + 2.228s\sqrt{\left(\dfrac{1}{7} + \dfrac{1}{5}\right)}.$$

For the data here,

$$\bar{t}_1 = 48.614, \quad s_1^2 = 2.2514$$
$$\bar{t}_2 = 48.380, \quad s_2^2 = 0.8370$$

so that

$$s^2 = \frac{(7-1)s_1^2 + (5-1)s_2^2}{(7+5-2)} = 1.6857$$

Thus the 95% confidence interval is:

$$-1.4595 < (\mu_1 - \mu_2) < 1.9281.$$

Note the considerable width of this confidence interval. In fact you would not be very surprised to discover that Runner 2 was intrinsically slower than Runner 1. You do not have sufficient data to find a very narrow band for the difference in underlying ability – though in practice the selection of one athlete over another for a team often depends on evidence of this type.

A GENERAL METHOD

Suppose that values of two random variables, X_1 and X_2, are measured on random samples of sizes n_1 and n_2. Let:

- X_1 and X_2 be distributed normally in the population with a common variance

- X_1 and X_2 be independent of each other in the population

- s^2 be the pooled-sample estimate of the common population variance of X_1 and X_2

- τ_a be the two-sided $a\%$ critical value for the t-distribution with $(n_1 + n_2 - 2)$ degrees of freedom.

Then a $(100 - a)\%$ confidence interval for the difference in the means μ_1 and μ_2 of X_1 and X_2 is given by:

$$(\bar{x}_1 - \bar{x}_2) - \tau_a s \sqrt{\frac{1}{n_1} + \frac{1}{n_2}} < (\mu_1 - \mu_2) < (\bar{x}_1 - \bar{x}_2) + \tau_a s \sqrt{\frac{1}{n_1} + \frac{1}{n_2}}.$$

EXERCISE 4B

1 The masses, in grams, of nine hens' eggs and eight ducks' eggs are recorded below.

Hens	42	47	45	41	48
	39	46	45	48	
Ducks	45	47	51	46	49
	53	53	48		

(i) Construct a 95% confidence interval for the difference in mean masses of hens' eggs and ducks' eggs.

(ii) State the assumptions you are making in constructing this confidence interval.

2 Nineteen pairs of brothers, where the elder was born two years before the younger, have their salaries at age 25 recorded (in thousands of pounds, to the nearest thousand pounds):

Salary of:

older brother	younger brother	older brother	younger brother
23	7	35	12
14	9	8	7
16	16	7	9
11	25	9	9
9	12	11	6
17	8	10	8
9	6	7	13
11	8	8	7
9	8	8	8
16	10		

(i) Use these data to determine a 90% confidence interval for the mean amount by which the elder brother's salary exceeds the younger's at age 25.

(ii) What assumptions are you making in constructing your confidence interval?

3 The table below shows the times, in seconds, that 13 rats took to run a maze on the first and second times they performed this task.

First run	Second run	First run	Second run
43	29	56	55
29	11	23	20
50	42	39	15
17	21	31	17
28	12	55	22
41	17	48	40
		36	19

Find a 98% confidence interval for the mean reduction in time between the first and second run.

4 A group of rowers and a group of chess players have their resting pulse rates measured. These data are shown below.

Rowers	65	73	71	80	61
	77	83	70	76	

Chess players	117	93	68	92	73
	102	85			

Construct a one-sided 95% confidence interval, giving an upper limit for the extent to which the mean rest pulse rate of chess players exceeds that of rowers.

5 The amount, p, of infestation of turnip fields by root borers, in grams of the pest per square metre, is measured in randomly chosen square metre areas on 33 turnip farms. Some of the farms have sprayed the crops with a new pesticide. The measurements are summarised in the table below.

	Using new pesticide	Not using new pesticide
Number of farms	14	19
Σp	8 831	16 573
Σp^2	5 692 287	14 908 662

Construct a 90% confidence interval for the difference in mean infestation between the sprayed and unsprayed crops.

1 Given an experiment which produces pairs of values (X, Y) of two random variables, where $(X - Y)$ is normally distributed, to construct a *confidence interval* for the mean difference

$$E[X] - E[Y]$$

using as data n pairs (x_i, y_i) which provide an independent random sample of values of $(X - Y)$:

- calculate the difference $(x_i - y_i)$ for each data pair

- find the mean $\bar{d}$ and unbiased variance estimate s^2 from the differences

- select the appropriate critical value τ of the t-distribution with $(n - 1)$ degrees of freedom.

- The confidence interval is

$$\bar{d} - \tau \frac{s}{\sqrt{n}} < E[X] - E[Y] < \bar{d} + \tau \frac{s}{\sqrt{n}}.$$

2 Given an experiment which produces pairs of values of two random variables X and Y, each of which is normally distributed with the same variance, to construct a *confidence interval* for the mean difference

$$E[X] - E[Y]$$

using as data m values x_i and n values y_j which constitute independent random samples of X and Y, respectively:

- find the means $\bar{x}, \bar{y}$ of the data sets

- calculate the unbiased estimate of the common variance s^2 from the pooled samples

- select the appropriate critical value τ of the t-distribution with $(m + n - 2)$ degrees of freedom.

- The confidence interval is

$$\bar{x} - \bar{y} - \tau s \sqrt{\frac{1}{m} + \frac{1}{n}} < E[X] - E[Y] < \bar{x} - \bar{y} + \tau s \sqrt{\frac{1}{m} + \frac{1}{n}}.$$

5 Large sample tests and confidence intervals

Experience isn't interesting till it begins to repeat itself – in fact, till it does that, it hardly is experience.

Elizabeth Bowen

In a charity trivia game, players must pay £1.50 for a card and then match the words on the card to their definitions.

> **MATCH THE WORD TO THE DEFINITION**
> **A:** *Peen* **B:** *Pigsney* **C:** *Noyade*
> **1:** A term of endearment
> **2:** The thin end of a hammer-head
> **3:** Mass execution by drowning

The prize to be won is £1.00 for each correct match.

How much can the charity expect to make per card and how predictable is the actual amount they will make per card, if the charity sells many cards?

On the card shown, the correct matches are A to 2, B to 1 and C to 3. This would win £3.00. There are three ways of winning £1.00 (A1, B2, C3; A2, B3, C1; A3, B1, C2). The two remaining possible matchings (A1, B3, C2; A3, B2, C1) win nothing.

Since the words are not well known, you can assume that players guess at random. For such a player, the probability distribution of the random variable W, the amount won per card, is

Amount won in £	Probability
0	$\dfrac{2}{6} = \dfrac{1}{3}$
1	$\dfrac{3}{6} = \dfrac{1}{2}$
3	$\dfrac{1}{6}$

This distribution is illustrated graphically in figure 5.1.

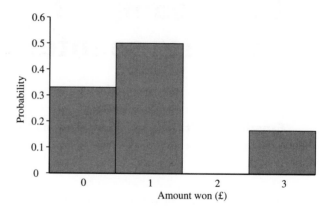

Figure 5.1

The expected win per card is therefore:

$$E[W] = 0 \times \frac{1}{3} + 1 \times \frac{1}{2} + 3 \times \frac{1}{6} = 1,$$

so that players who guess will lose on average 50p per card.

The variance of the win per card is:

$$\text{Var}[W] = 0^2 \times \frac{1}{3} + 1^2 \times \frac{1}{2} + 3^2 \times \frac{1}{6} - 1^2 = 1.$$

If a player buys four cards, what could his average win per card be and how likely is each possible value? His total win must be an integer between 0 and 12, so his average win (total divided by 4) must be a multiple of $\frac{1}{4}$ between 0 and 3.

P(player's average win per card over four cards = 0)

= P(zero matches on each of four cards)

$$= \left(\frac{1}{3}\right)^4 = \frac{1}{81}$$

P(player's average win per card over four cards = 0.25)

= P(one match on one card, zero on the other three)

$$= \binom{4}{1} \times \frac{1}{2} \times \left(\frac{1}{3}\right)^3 = \frac{2}{27}$$

P(player's average win per card over four cards = 0.5)

= P(one match on two cards, zero on the other two)

$$= \binom{4}{2} \times \left(\frac{1}{2}\right)^2 \times \left(\frac{1}{3}\right)^2 = \frac{1}{6}$$

$$P(\text{player's average win per card over four cards} = 0.75)$$

$$= P(\text{one match on three cards, zero on the other}$$

$$\text{three or three matches on one card, zero on the}$$

$$\text{other three})$$

$$= \binom{4}{3} \times \left(\frac{1}{2}\right)^3 \times \frac{1}{3} + \binom{4}{1} \times \frac{1}{6} \times \left(\frac{1}{3}\right)^3 = \frac{31}{162}$$

Calculations continue in this way, until the distribution tabulated and illustrated in figure 5.2 is built up.

Average amount won per card	Probability
0	$\frac{1}{81}$
0.25	$\frac{2}{27}$
0.5	$\frac{1}{6}$
0.75	$\frac{31}{162}$
1	$\frac{25}{144}$
1.25	$\frac{1}{6}$
1.5	$\frac{11}{108}$
1.75	$\frac{1}{18}$
2	$\frac{1}{24}$
2.25	$\frac{1}{162}$
2.5	$\frac{1}{108}$
2.75	0
3	$\frac{1}{1296}$

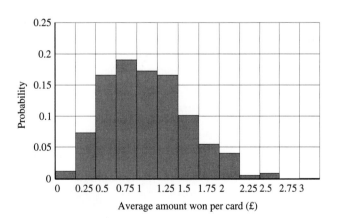

Figure 5.2

Imagine many people buying four cards, playing the game with each, and calculating their average win per card over the four cards. The distribution constructed above represents the theoretical proportions with which the different possible results would come up. You can think of each person as taking a sample of size four from the distribution of W and finding the mean of their sample. The distribution constructed above is that of $\overline{W}$, the random variable which represents the mean win per card in a sample of four cards.

How does the distribution of $\overline{W}$ compare with that of W?

- The two distributions have the same range; it is possible to win £3 or £0 on each card, hence have a mean win of £3 or £0 over four cards.

- Both distributions have the same expectation:

$$
\begin{aligned}
\mathrm{E}[\overline{W}] &= \mathrm{E}\left[\frac{1}{4}(W_1 + W_2 + W_3 + W_4)\right] \\
&= \frac{1}{4}(\mathrm{E}[W_1] + \mathrm{E}[W_2] + \mathrm{E}[W_3] + \mathrm{E}[W_4]) \\
&= 1.
\end{aligned}
$$

($W_1, W_2, \ldots$ are the random variables representing the amounts won on the 1st, 2nd, ... card.)

- The distribution of $\overline{W}$ is more peaked near to its mean than that of W: in fact, the variance of $\overline{W}$ is

$$
\begin{aligned}
\mathrm{Var}[\overline{W}] &= \mathrm{Var}\left[\frac{1}{4}(W_1 + W_2 + W_3 + W_4)\right] \\
&= \frac{1}{16}(\mathrm{Var}[W_1] + \mathrm{Var}[W_2] + \mathrm{Var}[W_3] + \mathrm{Var}[W_4]) \\
&= \frac{1}{4}
\end{aligned}
$$

so the standard deviation of $\overline{W}$ is half that of W.

- The distribution of $\overline{W}$ is more symmetrical and less skewed, than that of W.

You can repeat the calculations shown above (a computer is very useful) to find the distribution of the mean win per card in larger samples. For instance, the cases where each player buys 16 and 64 cards and calculates the mean win per card are illustrated in figure 5.3 and figure 5.4. (The tables are rather long.)

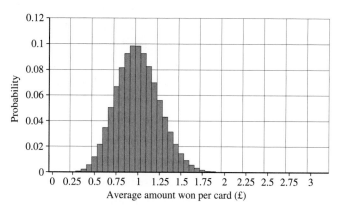

Figure 5.3 Mean win per card in a sample of 16 cards

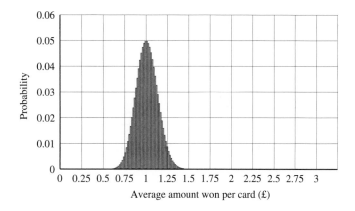

Figure 5.4 Mean win per card in a sample of 64 cards

You know the general results, for a sample of size n,

$$E[\overline{W}] = E[W]$$

$$Var[\overline{W}] = \frac{Var[W]}{n}$$

so that you would expect,

- for samples of size 16, $\overline{W}$ will have mean 1 and standard deviation $\frac{1}{4}$

- for samples of size 64, $\overline{W}$ will have mean 1 and standard deviation $\frac{1}{8}$.

This explains why the peak of the distribution of the sample mean remains near 1 whatever the sample size. It also explains why, although in principle the distributions of the sample mean for larger sample sizes continue to have the same range, from 0 to 3, the distribution becomes more peaked near to its mean as the sample size increases.

The final feature of the distribution of $\overline{W}$ noted above was that it is more symmetrical and less skewed, than that of W: this continues as the sample size increases. In fact a stronger statement can be made: for large sample sizes, the shape of the distribution appears to be approaching that of the normal distribution. Figure 5.5 and figure 5.6 show again the distributions of $\overline{W}$ with sample sizes 16 and 64: superimposed are the normal density curves with the same mean and variance as $\overline{W}$ in each case.

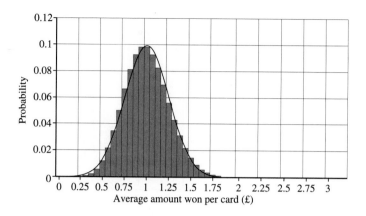

Figure 5.5

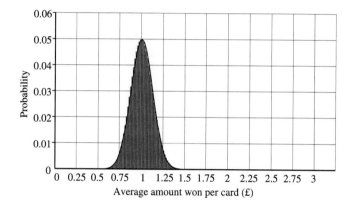

Figure 5.6

Visually, it appears that for a sample size of 64, the normal curve is an excellent fit. It is surprisingly good even for a sample size of 16. Of course the distribution of $\overline{W}$ remains discrete, while the normal distribution is continuous, so the two distributions do not become 'the same', but for the purposes of hypothesis testing and constructing confidence intervals, your interest in the distribution of $\overline{W}$ will be to enable you to calculate tail probabilities. For instance, you may wish to find its lower critical value at the 5% level, that is the value of a for which $P(\overline{W} < a) = 0.05$.

The table on page 91 shows part of the exact distribution (to 6 decimal places) of $\overline{W}$ for samples of size 64, together with cumulative probabilities.

a	$P(\overline{W} = a)$	$P(\overline{W} \leqslant a)$
0.125	0.000 000	0.000 000
0.140 625	0.000 000	0.000 000
0.156 25	0.000 000	0.000 000
0.171 875	0.000 000	0.000 000
0.187 5	0.000 000	0.000 000
0.203 125	0.000 000	0.000 000
0.218 75	0.000 000	0.000 000
0.234 375	0.000 000	0.000 000
0.25	0.000 000	0.000 000
0.265 625	0.000 000	0.000 000
0.281 25	0.000 000	0.000 000
0.296 875	0.000 000	0.000 000
0.312 5	0.000 000	0.000 000
0.328 125	0.000 000	0.000 000
0.343 75	0.000 000	0.000 000
0.359 375	0.000 000	0.000 000
0.375	0.000 000	0.000 000
0.390 625	0.000 000	0.000 001
0.406 25	0.000 001	0.000 001
0.421 875	0.000 001	0.000 003
0.437 5	0.000 002	0.000 005
0.453 125	0.000 004	0.000 008
0.468 75	0.000 006	0.000 014

a	$P(\overline{W} = a)$	$P(\overline{W} \leqslant a)$
0.484 375	0.000 010	0.000 024
0.5	0.000 017	0.000 041
0.515 625	0.000 028	0.000 069
0.531 25	0.000 044	0.000 113
0.546 875	0.000 070	0.000 184
0.562 5	0.000 110	0.000 294
0.578 125	0.000 169	0.000 462
0.593 75	0.000 255	0.000 718
0.609 375	0.000 380	0.001 098
0.625	0.000 557	0.001 654
0.640 625	0.000 804	0.002 458
0.656 25	0.001 142	0.003 600
0.671 875	0.001 597	0.005 196
0.687 5	0.002 199	0.007 395
0.703 125	0.002 980	0.010 375
0.718 75	0.003 978	0.014 353
0.734 375	0.005 227	0.019 580
0.75	0.006 762	0.026 342
0.765 625	0.008 612	0.034 954
0.781 25	0.010 799	0.045 754
0.796 875	0.013 332	0.059 085
0.812 5	0.016 203	0.075 288
0.828 125	0.019 388	0.094 676

From this the critical value can be read off:

$$P(\overline{W} \leqslant 0.78125) \leqslant 0.05 \text{ (and } P(\overline{W} \leqslant 0.796875) > 0.05).$$

Using the normal approximation $\overline{W} \approx N\left(1, \frac{1}{64}\right)$, the critical value is given by (using 1.645 as the 5% one-tailed critical value for the standard normal distribution)

$$1 - 1.645 \times \frac{1}{8} = 0.794\,375$$

which gives precisely the same set of samples in the critical region as the exact critical value, since the sample value of $\overline{W}$ can take only the values listed in the table above and

$$0.781\,25 < 0.794\,375 < 0.796\,875.$$

This example illustrates the claim of *The Central Limit Theorem*:

Assume random samples of size n are drawn from a distribution, which need not be normal, with mean μ and variance σ^2.

For large n, the probability of the sample mean being in a given range can be well-approximated by the corresponding area under the normal density function $N\left(\mu, \dfrac{\sigma^2}{n}\right)$.

The organisers of the trivia game could ask how much money they are likely to make if they sell more than 2500 cards (and none of their customers has an absurdly large vocabulary). The number of cards they sell is their sample size so, if they sell more than 2500 cards, the average win per card has variance less than $\frac{1}{2500}$ and therefore a standard deviation less than $\frac{1}{50}$. The probability that the average win per card will be outside the range

$$1 - 2.576 \times \frac{1}{50} \approx 0.95 \text{ to } 1 + 2.576 \times \frac{1}{50} \approx 1.05$$

is less than 1%, since 2.576 is the two-tailed 1% critical value for the standard normal distribution.

The organisers' profit is therefore very unlikely to be outside the range of 45p to 55p per card they sell and, if they sell a lot more than 2500 cards, this range could be narrowed.

Tests and confidence intervals when the variances are known

In studying unpaired t-tests, you had to make the assumption that the variance in the population of the random variable you were sampling was the same in both conditions. You then estimated this common variance from the two samples. However, there are some situations in which you know the variance of the whole population and you can use this information in a hypothesis test or in constructing confidence intervals.

For instance, it may be that, before the ability of the maths class to estimate one minute was tested (see page 57), extensive tests were conducted which determined that, in the school population as a whole, students' estimated minutes are normally distributed and have a standard deviation of 7.42 seconds. You are testing the hypotheses:

H_0: There is no difference between the mean of people's estimates of one minute before and after lunch.

H_1: After lunch, the mean of people's estimates of one minute tends to be shorter than before lunch.

But you can now make the assumption that people's estimates of one minute are normally distributed with standard deviation 7.42.

The null hypothesis implies that before-lunch and after-lunch estimates have distributions $N(\mu, 7.42^2)$ where μ is the common mean asserted by the null hypothesis. With this assumption, the mean of the 24 before-lunch estimates has distribution:

$$\overline{X} \sim N\left(\mu, \frac{7.42^2}{24}\right)$$

and the mean of the 22 after-lunch estimates has distribution:

$$\overline{Y} \sim N\left(\mu, \frac{7.42^2}{22}\right)$$

The distribution of the difference of the two sample means is therefore:

$$\overline{X} - \overline{Y} \sim N\left(0, \frac{7.42^2}{24} + \frac{7.42^2}{22}\right)$$

In *Statistics 3* you considered hypothesis tests with the normal distribution: if X has distribution $N(0, \sigma^2)$, where the variance σ^2 is known, then the test statistic $\dfrac{X}{\sigma}$ has the standard normal distribution, $N(0, 1)$. The test statistic here is:

$$\frac{\overline{X} - \overline{Y}}{\sqrt{\dfrac{7.42^2}{24} + \dfrac{7.42^2}{22}}} = \frac{\overline{X} - \overline{Y}}{7.42\sqrt{\dfrac{1}{24} + \dfrac{1}{22}}}$$

With the data used in the original example, $\bar{x} = 51.542$ and $\bar{y} = 47.909$, so the test statistic has value:

$$\frac{51.542 - 47.909}{7.42\sqrt{\dfrac{1}{24} + \dfrac{1}{22}}} = 1.659$$

The critical region for a one-tailed test at the 5% significance level for the standard normal distribution is $z > 1.645$. In this case, since $1.659 > 1.645$, you reject the null hypothesis and accept the alternative, that the after-lunch times are shorter than the before-lunch times.

Different known variances for the two samples

Alternatively, you might know separately the variances of the populations from which each sample was drawn, where these need not be the same.

Suppose there are two machines in a factory. The first is a high-accuracy machine, which produces bolts with radii which are normally distributed with standard deviation 0.052 mm. The second is a lower-accuracy machine, producing washers with internal radii which are normally distributed with standard deviation 0.172 mm. Both machines are adjustable to produce components with different radii, but today they are supposed to be set so that the high-accuracy machine produces bolts with radii 2 mm smaller than the internal radii of the washers produced by the low-accuracy machine.

To check whether the setting is correct, a sample of components is taken from each machine, and the radius of each measured. The results are shown in the following table.

Radii of bolts from high accuracy machine (mm)
8.42,   8.21,   8.29,   8.31,   8.25,   8.38,   8.29

Internal radii of washers from low-accuracy machine (mm)
10.32,   10.12,   9.98,   10.09,   10.57,   10.49,   10.10,   10.28,   10.35

You are testing the hypotheses:

H_0: The mean radius of the bolts being produced is 2 mm less than the mean internal radius of the washers being produced.

H_1: The mean radius of the bolts and the mean internal radius of the washers being produced do not differ by 2 mm.

You can assume that the radii of the components being produced by each machine are normally distributed with the standard deviations given above.

If X_W denotes the internal radius of a washer, and X_B the radius of a bolt, what is the distribution of the sample statistic $\bar{X}_W - \bar{X}_B$?

With the assumptions stated, the mean internal radius of nine washers from the low-accuracy machine has distribution:

$$\overline{X}_W \sim N\left(\mu_W, \frac{0.172^2}{9}\right)$$

where μ_W is the mean internal radius of the washers.

Similarly, the mean radius of seven bolts from the high-accuracy machine has distribution:

$$\overline{X}_B \sim N\left(\mu_B, \frac{0.052^2}{7}\right)$$

where μ_B is the mean radius of the bolts.

The distribution of the difference of the two sample means is therefore:

$$\overline{X}_W - \overline{X}_B \sim N\left(\mu_W - \mu_B, \frac{0.172^2}{9} + \frac{0.052^2}{7}\right).$$

The null hypothesis then states that $\mu_W - \mu_B = 2$ and so, if the null hypothesis is true:

$$\overline{X}_W - \overline{X}_B \sim N\left(2, \frac{0.172^2}{9} + \frac{0.052^2}{7}\right).$$

Therefore, the test statistic:

$$\frac{\overline{X}_W - \overline{X}_B - 2}{\sqrt{\frac{0.172^2}{9} + \frac{0.052^2}{7}}}$$

has a standard normal distribution.

In this case $\overline{x}_W = 10.256$ and $\overline{x}_B = 8.307$, so the test statistic is:

$$\frac{10.256 - 8.307 - 2}{\sqrt{\frac{0.172^2}{9} + \frac{0.052^2}{7}}} = -0.84$$

The critical region for a two-tailed test with the standard normal distribution at the 5% significance level is $z > 1.96$ or $z < -1.96$, and so here, since $-1.96 < -0.84 < 1.96$, you accept the null hypothesis that the machines are correctly set.

You can use the same data to construct a 95% confidence interval for the difference in mean radii being produced by the two machines. You saw above that the distribution of the difference of the two sample means is:

$$\overline{X}_W - \overline{X}_B \sim N\left(\mu_W - \mu_B, \frac{0.172^2}{9} + \frac{0.052^2}{7}\right)$$

so that:

$$P\left(-1.96 < \frac{(\overline{X}_W - \overline{X}_B) - (\mu_W - \mu_B)}{\sqrt{\dfrac{0.172^2}{9} + \dfrac{0.052^2}{7}}} < 1.96\right) = 0.95$$

since 1.96 is the two-tailed 5% critical value for the standard normal distribution. The confidence interval is therefore:

$$(\overline{x}_W - \overline{x}_B) - 1.96\sqrt{\frac{0.172^2}{9} + \frac{0.052^2}{7}} < (\mu_W - \mu_B) < (\overline{x}_W - \overline{x}_B) + 1.96\sqrt{\frac{0.172^2}{9} + \frac{0.052^2}{7}}.$$

With the values $\overline{x}_W = 10.256$ and $\overline{x}_B = 8.307$, this is:

$$1.83 < \mu_W - \mu_B < 2.07.$$

Summary: known variances

The null hypothesis is:

H_0: The difference between the means of the random variable in the two conditions is $(\mu_1 - \mu_2)$.

If:

- the random variable, X, has a normal distribution in each condition

- $\overline{X}_1$ and $\overline{X}_2$ are the means of the samples in the two conditions

- n_1 and n_2 are the sizes of these samples

- σ_1 and σ_2 are the known standard deviations of the random variables X_1 and X_2,

then the test statistic is:

$$\frac{\overline{X}_1 - \overline{X}_2 - (\mu_1 - \mu_2)}{\sqrt{\dfrac{\sigma_1^2}{n_1} + \dfrac{\sigma_2^2}{n_2}}}$$

which has a standard normal distribution N(0, 1).

In the same situation, a $(100 - a)\%$ confidence interval for the difference of the means in the two conditions is:

$$(\overline{x}_1 - \overline{x}_2) - z_a\sqrt{\frac{\sigma_1^2}{n_1} + \frac{\sigma_2^2}{n_2}} < \mu_1 - \mu_2 < (\overline{x}_1 - \overline{x}_2) + z_a\sqrt{\frac{\sigma_1^2}{n_1} + \frac{\sigma_2^2}{n_2}}$$

where z_a is the two-sided $a\%$ critical value for the standard normal distribution.

If the variances in the two conditions are known and equal, each distribution having standard deviation σ, then the test statistic simplifies to:

$$\frac{(\overline{X}_1 - \overline{X}_2) - (\mu_1 - \mu_2)}{\sigma\sqrt{\dfrac{1}{n_1} + \dfrac{1}{n_2}}}$$

and the confidence interval to:

$$(\overline{x}_1 - \overline{x}_2) - z_a\sigma\sqrt{\frac{1}{n_1} + \frac{1}{n_2}} < \mu_1 - \mu_2 < (\overline{x}_1 - \overline{x}_2) + z_a\sigma\sqrt{\frac{1}{n_1} + \frac{1}{n_2}}.$$

You can do the same sort of thing for the paired test, with null hypothesis:

H_0: The difference between the values of the random variable in the two conditions has mean d.

If:

- the difference, X, between the pairs of values of the random variable in the two conditions has a normal distribution

- $\overline{X}$ is the mean of X

- σ is the known standard deviation of X

- n is the size of the sample

then the test statistic is $\dfrac{\overline{X} - d}{\frac{\sigma}{\sqrt{n}}}$ which has a standard normal distribution, N(0, 1).

In the same situation, a $(100 - a)\%$ confidence interval for the mean difference between the two conditions is:

$$\overline{X} - z_a\frac{\sigma}{\sqrt{n}} < \mu_1 - \mu_2 < \overline{X} + z_a\frac{\sigma}{\sqrt{n}}$$

where z_a is the two-sided $a\%$ critical value for the standard normal distribution.

Note

This situation – where you know the variance of the difference between the random variables in the two conditions – is very unlikely to arise in practice. Knowing the variance of the random variable in each condition separately is not sufficient, as the point of pairing is that the variable is not independently measured in the two conditions. Thus in a paired test, if X_1 and X_2 represent the random variable in the two conditions, $\mathrm{Var}[X_1 - X_2] \neq \mathrm{Var}[X_1] + \mathrm{Var}[X_2]$.

It may have occurred to you that, in fact, the examples you have considered so far in this chapter were rather contrived, and that is not surprising: it is difficult to think of circumstances where the tests described here actually apply – why should you know the variances of the distributions, but not their means? This does not mean that you have been wasting your time, however: the importance of this technique is seen in the next section.

5

Tests with large samples

In Chapters 4 and 5, when you used the t-distribution for testing hypotheses about differences between means and for constructing confidence intervals for the difference between means, you had to make the assumptions:

1 the underlying variables are normally distributed

2 the variables have a common variance.

In many situations where you want to test hypotheses or construct confidence intervals, these assumptions do not hold. Fortunately, provided that a large sample is available, these assumptions are not essential.

1 The central limit theorem says that, for large sample sizes, even when a variable does not have a normal distribution, its sample mean is approximately normally distributed.

2 In general, the larger the sample being used, the smaller the error made in assuming that the population has exactly the variance given by its sample estimate.

The tests discussed so far in this chapter, which assume normally-distributed variables with known variances, will therefore all give sensible results for large samples, even where the underlying variables are not normally distributed and the variances are not known, but must be estimated from the samples. As a rule of thumb, sample sizes of 50 or so are large enough for this approximation to be reasonable – although this will obviously depend on how non-normal the underlying distributions are. In many situations the approximation will be justified with substantially smaller samples.

Even where it cannot be assumed that the underlying variables have a common variance, but where large samples are available, separate population variances can be estimated and used as if they were the actual population variances.

EXAMPLE 5.1

Smartos tubes are filled by two different machines. A sample of tubes filled by each machine is taken and the Smartos in each tube are counted. The results are given in the table below.

Number of Smartos per tube	39	40	41	42	43	Total
Frequency in machine A sample	17	23	35	31	27	133
Frequency in machine B sample	21	18	41	39	19	138

Assuming a common variance for the numbers of Smartos per tube produced by the machines, construct a 98% confidence interval for the difference in the mean number of Smartos in the tubes produced by the two machines.

SOLUTION

Here, you certainly cannot assume that the distribution of the number of Smartos per tube is normal since it is a discrete variable. Nonetheless, because the sample size is large, and you are told that there is a common variance in the two conditions, you can assume that the statistic:

$$\frac{(\overline{X}_A - \overline{X}_B) - (\mu_A - \mu_B)}{S\sqrt{\dfrac{1}{n_A} + \dfrac{1}{n_B}}}$$

has approximately a standard normal distribution. ($\overline{X}_A$ and $\overline{X}_B$ are the means of the numbers of Smartos per tube in the sample from each machine, S^2 is the pooled-sample estimator of the common variance, n_A and n_B are the sizes of each sample, and μ_A and μ_B are the true mean numbers of Smartos produced by the two machines.)

The two-tailed 2% critical value for the standard normal distribution is 2.326, so that for approximately 98% of samples:

$$-2.326 < \frac{(\overline{x}_A - \overline{x}_B) - (\mu_A - \mu_B)}{s\sqrt{\dfrac{1}{n_A} + \dfrac{1}{n_B}}} < 2.326.$$

That is, an approximate 98% confidence interval for the difference between the true mean numbers is:

$$(\overline{x}_A - \overline{x}_B) - 2.326s\sqrt{\frac{1}{n_A} + \frac{1}{n_B}} < \mu_A - \mu_B < (\overline{x}_A - \overline{x}_B) + 2.326s\sqrt{\frac{1}{n_A} + \frac{1}{n_B}}.$$

Here:

$$n_A = 133, \ \overline{x}_A = 41.2105, \ s_A = 1.3030$$

$$n_B = 138, \ \overline{x}_B = 41.1232, \ s_B = 1.2525$$

so that:

$$s = \sqrt{\frac{132 \times 1.3030^2 + 137 \times 1.2525^2}{133 + 138 - 2}} = 1.2775$$

and so the confidence interval is:

$$-0.2738 < \mu_A - \mu_B < 0.4484.$$

General procedure for large samples

In an experiment with an unpaired design a test of the hypothesis:

H_0: The difference between the means in the two conditions is δ

uses data from two samples, one taken in each condition. These data are summarised by:

- $\overline{X}_1$ and $\overline{X}_2$, the means of the samples in the two conditions
- n_1 and n_2, the sizes of these samples
- S_1^2 and S_2^2 the sample estimators of the variances in the two conditions.

The test statistic is:

$$\frac{\overline{X}_1 - \overline{X}_2 - \delta}{\sqrt{\dfrac{S_1^2}{n_1} + \dfrac{S_2^2}{n_2}}}$$

which has approximately a standard normal distribution, if the sample size is large.

In the same situation, a $(100 - a)\%$ confidence interval for the difference of the means in the two conditions is:

$$(\overline{x}_1 - \overline{x}_2) - z_a\sqrt{\frac{s_1^2}{n_1} + \frac{s_2^2}{n_2}} < \mu_1 - \mu_2 < (\overline{x}_1 - \overline{x}_2) + z_a\sqrt{\frac{s_1^2}{n_1} + \frac{s_2^2}{n_2}}$$

where z_a is the two-sided $a\%$ critical value for the standard normal distribution.

SPECIAL CASES

1 If you know that the variances in the two conditions have the values of σ_1^2 and σ_2^2 these values should be used instead of the sample estimates s_1^2 and s_2^2 in the test statistic and the confidence limits.

2 If you believe that the variances are the same in the two conditions you use the statistic:

$$\frac{\overline{X}_1 - \overline{X}_2 - \delta}{S\sqrt{\dfrac{1}{n_1} + \dfrac{1}{n_2}}}$$

where $S^2 = \dfrac{(n_1 - 1)S_1^2 + (n_2 - 1)S_2^2}{(n_1 + n_2 - 2)}$ is the pooled sample variance estimate.

Similarly, the confidence interval in this case is:

$$(\overline{x}_1 - \overline{x}_2) - z_a s\sqrt{\frac{1}{n_1} + \frac{1}{n_2}} < \mu_1 - \mu_2 < (\overline{x}_1 - \overline{x}_2) + z_a s\sqrt{\frac{1}{n_1} + \frac{1}{n_2}}.$$

In an experiment with a paired design a test of the hypothesis:

H_0: The mean difference between variables in the two conditions is δ,

uses data summarised by:

- $\overline{D}$, the mean of the sample differences between the two conditions

- n, the size of the samples

- S^2, the sample estimator of the variance of the differences between the two conditions.

The test statistic is $\dfrac{\overline{D} - \delta}{\dfrac{S}{\sqrt{n}}}$, which has an approximately standard normal distribution, if the sample size is large.

In the same situation, a $(100 - a)\%$ confidence interval for the difference of the means in the two conditions is

$$\overline{d} - z_a \frac{s}{\sqrt{n}} < \delta < \overline{d} + z_a \frac{s}{\sqrt{n}}$$

where z_a is the two-sided $a\%$ critical value for the standard normal distribution.

SPECIAL CASE

If you know that the variance of the difference between the two conditions has the value σ^2 this value should be used instead of the sample estimate s^2 in the test statistic and the confidence limits.

EXAMPLE 5.2

A test is devised to decide whether eggs in one supermarket are fresher than those in another. In store A, a carton of each of the six grades of egg on sale is examined every day for a month and, from the date code on each carton, the value of random variable X_A, the number of days since packing, is determined. This gives 186 observations of values X_A. In store B, only five grades of egg are on sale: a carton of each is examined every day for a month and the value of random variable X_B, the number of days since packing, is determined. This gives 155 observations of values of X_B.

The hypotheses are:

H_0: The mean number of days since packing is the same in both stores.

H_1: The mean number of days since packing is not the same in both stores.

The data are summarised by the values:

$$\Sigma x_A = 1722 \quad \Sigma x_A^2 = 17\ 854 \quad n_A = 186$$
$$\Sigma x_B = 1061 \quad \Sigma x_B^2 = 8533 \quad n_B = 155$$

SOLUTION

In this case, you should not expect the distributions of X_A and X_B to be normal because they are discrete. Nor should you necessarily want to assume that they have the same variance. However, because the samples are large, you are justified in assuming that, if the null hypothesis is true, the test statistic:

$$\frac{\overline{X}_A - \overline{X}_B}{\sqrt{\dfrac{S_A^2}{186} + \dfrac{S_B^2}{155}}}$$

has approximately a standard normal distribution.

From the data given:

$$\overline{x}_A = \frac{1722}{186} = 9.258$$

$$s_A = \sqrt{\frac{17\,854 - 186 \times 9.258^2}{185}} = 3.215$$

$$\overline{x}_B = \frac{1061}{155} = 6.845$$

$$s_B = \sqrt{\frac{8533 - 155 \times 6.845^2}{154}} = 2.872$$

so the test statistic is:

$$\frac{9.258 - 6.845}{\sqrt{\dfrac{3.215^2}{186} + \dfrac{2.872^2}{155}}} = 7.316.$$

The two-tailed critical region for the standard normal distribution at the 5% significance level is $z > 1.96$ or $z < -1.96$, so for this data, since $7.316 > 1.96$, you can reject the null hypothesis and accept that the eggs in one supermarket are fresher than those in the other.

1 There is a misty patch of Irish turf which is supposed to produce multi-leaved clovers with greater regularity than elsewhere. To test this hypothesis, clover stems were collected from this patch of turf, and from a large selection of other locations, and their leaves counted. The frequencies with which each number of leaves arose in the two conditions are shown in the table below.

Sample from the misty Irish turf

Number of leaves	3	4	5	6
Frequency	277	50	12	2

Sample from elsewhere

Number of leaves	3	4	5	6
Frequency	433	68	7	3

(i) Use these data to test the hypothesis that the spot described is special.

(ii) What assumptions are you making?

2 It has been suggested that, although the number of eggs laid by female hedgesparrows varies widely among birds, individual birds tend to lay more eggs each year that they breed. A sample of 162 female birds have their nests examined in two successive years, and the change in the number of eggs is recorded. The frequency of each change is recorded in the table below.

Change in number of eggs	−2	−1	0	1	2
Frequency	4	27	77	43	11

Use these data to construct a confidence interval for the mean increase in numbers of eggs per year, stating clearly the assumptions you are making.

3 A bank is deciding whether to introduce a new system for their cashiers. It is only worth the expense if it will reduce the average waiting time by a minute or more. They survey the waiting times for a sample of their customers in a branch and then, after introducing a trial run of the new system in the same branch, resurvey the waiting times. The results of the two surveys are shown in the table below.

	Freq	uency
Minutes in queue	With old system	With new system
0	33	49
1	57	58
2	44	44
3	35	24
4	17	12
5	13	2
6	10	4
7	6	1
8	5	0
9	5	0
10	1	1
11	4	1
12	3	0
13	3	0

Test the hypothesis that the waiting times have been reduced by at least one minute, on average, under the new system, stating the assumptions you are making.

4 Two very long shafts fitted to a turbine need to be very accurately the same length, but they cannot be moved to compare them, and are difficult to measure. The engineer whose job it is to check the lengths adopts the following procedure: ten separate measurements are made of each shaft, by a process of which the result is a normal variable with mean equal to the true length of the shaft and with standard deviation 3 hundredths of a millimetre.

The measurements made in one check are listed in the table below. (Lengths are given as hundredths of a millimetre from the nominal length of the shaft.)

First shaft	+6	+2	+7	−1	+4
Second shaft	−1	+3	+3	+5	−2

First shaft	−3	+3	+2	+7	+7
Second shaft	−4	+4	+2	−3	+0

Test the hypothesis that the two shafts have the same length.

5 The Governor of the Bank of England is attempting to keep the UK discount rate $\frac{1}{4}\%$ above the Lombard rate. The table below shows the frequency with which different combinations of rates are observed in the market at 12 noon each day over a period of four months.

UK discount rate (%)

Lombard rate (%)	5	5.25	5.5	5.75	6	6.25	6.5	6.75	7
5	1	12	7	1					
5.25	5	7	9	2					
5.5		2	2	15	5	4	1		
5.75				4	4	10			
6					3	8	2		
6.25				1		3	4		
6.5						1	5	4	

(i) Use these data to test the hypothesis that the UK rate is on average $\frac{1}{4}\%$ above the Lombard rate.

(ii) What assumptions are you making? How can this situation be viewed as a random sample from a particular distribution?

6 In an attempt to redesign a combustion chamber, it is necessary to find the difference between the maximum inside and outside temperatures of the casing. The combustion process is rather variable from one ignition to another: in fact the variance of the maximum temperature, in °C, inside the chamber is 3940 and the variance of the temperature of the outer casing is 2710.

Construct a confidence interval for the difference between mean inside and outside temperatures using the data below from a series of experimental ignitions.

Nine ignitions:	6870	6940	7010
maximum	6960	6890	6940
inside temp (°C)	6950	6920	6920
Six ignitions:	6710	6730	6680
maximum	6670	6680	6620
outside temp (°C)			

7 A new machine for packing matches into boxes has been delivered. It is supposed to be more accurate – that is, the variation in the number of matches it inserts into each box is less, but the factory manager also wants to check that the new machine is not extravagantly putting a higher average number of matches into each box. She takes a sample of boxes produced by the old and new machines and counts the frequency with which each number of matches arises. The results are given in the table.

Number of matches	40	41	42	43	44
Frequency with old machine	96	42	17	9	2
Frequency with new machine	65	107	19	0	0

Test the hypothesis that the two machines have equal means, stating the assumptions you are using.

8 In an attempt to decide whether a new feeding regime increases the average weight of young salmon in a fish farm, a sample of 165 fish fed in the usual way is weighed. The weights, x_i, in grams, of these fish are summarised by the figures:

$$\sum_{i=1}^{165} x_i = 11\,774 \qquad \sum_{i=1}^{165} x_i^2 = 872\,308$$

A sample of 74 fish is bred with the new feeding regime and their weights, y_i, in grams, are summarised by:

$$\sum_{i=1}^{74} y_i = 5491 \qquad \sum_{i=1}^{74} y_i^2 = 409\,272$$

(i) Test the hypothesis that the average weight of the salmon is higher under the new regime:

(a) assuming that the variance of their weights is unchanged

(b) estimating the variance separately in each condition.

(ii) Comment on your results.

9 In an experiment to determine whether presenting a list of words alphabetically or in random order causes them to be remembered more easily, two groups of subjects are given such lists of 30 words to study for one minute. After a distracting task, the subjects are then asked to recall as many words as possible in one minute. The numbers of words recalled in each condition are as follows.

Words recalled	14	15	16	17	18	19	20	21
Frequency: random list	0	0	0	1	10	12	34	78
Frequency: alphabetical list	3	2	23	57	17	21	19	33

Words recalled	22	23	24	25	26	27	28
Frequency: random list	55	22	29	11	2	1	2
Frequency: alphabetical list	48	26	19	4	8	0	0

(*Note:* The bimodal nature of the distribution with an alphabetical list is because subjects use two different strategies for recall in this case.)

(i) Test the hypothesis that there is no difference on average in the number of words recalled in the two conditions. State the assumptions that you are making.

(ii) Suggest a paired design for testing this hypothesis.

10 A new golf club is advertised with the claim 'add at least 20 metres to your drive', and a golfers' magazine is testing this claim.

It finds that the new club does enable its sample of golfers to hit further on average: the 120 golfers who used the new club on the test range hit the ball 367 metres on average, while the 142 golfers who used the old club hit only 353 metres on average. The new club was considerably more erratic in its performance, though. The sample estimates of the standard deviation of the length of the drives were 46 metres with the new club, and 27 metres with the old.

(i) Use these data to test the hypotheses:

(a) The new club enables golfers to drive further, on average, than the old.

(b) The new club enables golfers to drive at least 20 metres further, on average, than the old.

(ii) Construct a confidence interval for the increase in mean distance driven with the new club.

The remainder of this exercise gives practice in the techniques introduced throughout Chapters 3, 4 and 5.

11 Two friends, Mary and Susan, are planning to buy a carpet for the living room in their flat. They therefore measure the length and width of their room. The errors in their measurements may be taken as having normal distributions. Mary's distribution has mean 0 and standard deviation 3, Susan's has mean 0 and standard deviation 2 (all units are cm).

Mary measures the length of the room on four different occasions, obtaining measurements of 425 cm, 427 cm, 422 cm and 426 cm. Susan measures the width of the room once, obtaining a measurement of 366 cm.

Giving your answers in cm correct to one decimal place, find

(i) a one-sided 95% confidence interval giving a lower bound for the length of the room

(ii) a symmetrical two-sided 99% confidence interval for the width of the room.

One of Susan's friends, calling her calculated interval (l, u), states 'there is probability 99% that the width of the room is between l and u'. Explain why this interpretation is wrong. What is the correct interpretation?

12 For reasons of economy, the manufacturers of an electrical appliance wished to make an adjustment to one of its components. Before finally deciding whether to do so, the effect of the adjustment on the resistance of the component was assessed.

(i) At one factory, the resistances of ten such components selected at random were measured both before and after the adjustment. The results were as follows:

Component	Resistance (ohms) before adjustment	Resistance (ohms) after adjustment
1	37.7	41.0
2	42.1	47.8
3	44.2	44.9
4	35.2	39.6
5	38.6	45.8
6	43.2	45.9
7	47.3	49.6
8	35.9	38.7
9	43.7	44.5
10	42.4	49.1

Test at the 5% significance level whether there is a difference in mean resistance due to the adjustment.

(ii) At another factory, twenty of the components were selected at random and ten of these were chosen at random to receive the adjustment. The resistances of the components were measured, with the following results:

Unadjusted component's resistance (ohms), x	Adjusted component's resistance (ohms), y
42.3	49.2
35.3	44.8
44.3	39.5
42.0	47.7
37.6	39.9
43.6	44.4
35.8	49.3
47.6	38.8
43.1	45.8
38.5	45.7

(For information: $\Sigma x = 410.1$, $\Sigma x^2 = 16\,963.85$, $\Sigma y = 445.1$, $\Sigma y^2 = 19\,948.65$)

Test at the 5% significance level whether there is a difference between the mean resistances in each group, stating carefully the assumption you make about the underlying variances.

(iii) Explain clearly which of the two analyses gives better information and why.

13 Investigations are being made of the time taken to bring water to the boil in a large urn in a cafeteria. It is known that this time varies somewhat and that the variations may be accounted for by taking the boiling time to be normally distributed.

(i) The boiling times in minutes on ten randomly chosen occasions were 20.2, 17.8, 23.6, 21.1, 19.4, 19.6, 20.9, 20.0, 18.9, 20.3. Find a two-sided symmetrical 95% confidence interval for the true mean boiling time.

(ii) A second urn is acquired, for which the boiling time may again be taken as normally distributed but with a possibly different mean. A random sample of 20 boiling times for the second urn is found to have mean 19.33 minutes. Information from the manufacturers states that the *true* standard deviation of boiling time is 0.9 minutes for both urns. Assuming this is indeed correct, examine at the 5% level of significance whether the true mean boiling times for the two urns differ.

(iii) Suppose instead that the boiling times for the urns may not be taken as normally distributed. State the name of, and *briefly* describe, a statistical procedure that could be used to examine whether the two urns have, overall, the same boiling times.

14 Two personal computers are being compared with respect to their performances in running typical jobs. Eight typical jobs selected at random are run on each computer. The table shows the values of a composite unit of performance for each job on each computer.

Job	Computer A	Computer B
1	214	203
2	198	202
3	222	216
4	206	218
5	194	185
6	236	224
7	219	213
8	210	212

It is desired to examine whether the mean performance for typical jobs is the same for each computer.

(i) State formally the null and alternative hypotheses that are being tested.

(ii) State an appropriate assumption concerning the underlying normality.

(iii) Carry out the test, using a 1% level of significance.

(iv) Provide a symmetrical two-sided 95% confidence interval for the difference between the mean performance times.

15 The central business district of a town is served by two railway stations A and B. Part of a study is to examine whether the mean daily number of passengers arriving at station A during the morning peak period is the same as the corresponding average at station B. Counts were taken at station A for a random sample of 8 working days and at station B for a separate random sample of 12 working days, with the following results for the numbers of passengers arriving during the morning peak period.

Station A	1486	1529	1512	1540
	1506	1464	1495	1502
Station B	1475	1497	1460	1478
	1520	1473	1449	1480
	1503	1462	1474	1486

(i) State formally the null and alternative hypotheses that are to be tested.

(ii) State an appropriate assumption concerning underlying normality.

(iii) State a further necessary assumption concerning the underlying distributions.

(iv) Carry out the test, using a 5% level of significance.

(v) Suppose that, in a test situation such as this, the *true* variances of the underlying distributions were known. Outline *briefly* how the test would be conducted.

16 A liquid product is sold in containers. The containers are filled by a machine. The volumes of liquid (in ml) in a random sample of six containers were found to be

497.8   501.4   500.2   500.8   498.3   500.0.

After an overhaul of the machine, the volumes (in ml) in a random sample of 11 containers were found to be

501.1 499.6 500.3 500.9 498.7 502.1

500.4 499.7 501.0 500.1 499.3.

It is desired to examine whether the average volume of liquid delivered to a container by the machine is the same after overhaul as it was before.

(i) State the assumptions that are necessary for the use of the customary t-test.

(ii) State formally the null and alternative hypotheses that are to be tested.

(iii) Carry out the t-test, using a 5% level of significance.

(iv) Discuss briefly which of the assumptions in **(i)** is the least likely to be valid in practice and why.

17 A railway station has a telephone enquiry office. The length of time, in minutes, taken to deal with any caller's enquiry is independent of that for all other callers and is modelled by the continuous random variable X with probability density function

$$f(x) = \frac{1}{4}xe^{-x/2} \qquad 0 \leqslant x < \infty.$$

(i) Show that the mean of X is 4.

(*Reminder*: the limit of $x^m e^{-x}$ as $x \to \infty$ is zero.)

(ii) You are now given that the variance of X is 8. State the mean and variance of T, the combined time for dealing with eight callers.

(iii) Explain why a normal random variable will provide a good approximation to the distribution of T.

(iv) An attempt is made to improve the modelling. The detailed form of the X variable is discarded, though it is still believed appropriate to use a random

variable whose variance is twice the mean. Denoting this mean by θ (and therefore the variance by 2θ), write down the parameters of the normal distribution that is now to be used as an approximation to the distribution of T. Deduce that, according to this distribution,

$$P\left(\frac{T - 8\theta}{4\sqrt{\theta}} < 1.645\right) = 0.95.$$

(v) The combined time for dealing with eight callers is measured once and found to be 25 minutes. Show that, using the distribution in **(iv)**, the lower limit of a one-sided 95% confidence interval for θ is a solution of the equation

$$64\theta^2 - 443.2964\theta + 625 = 0$$

What does the other solution of this equation represent?

18 An inspector is examining the lengths of time taken to complete various routine tasks by employees who have been trained in two different ways. He wants to examine whether the two methods lead, overall, to the same times. Ten different tasks have been prepared. Each task is undertaken by a randomly selected employee who has been trained by method A and by a randomly selected employee who has been trained by method B. The times to completion, in minutes, are shown in the table.

Task	Time taken A employee	Time taken B employee
1	33.4	27.1
2	41.0	42.0
3	26.8	23.0
4	37.2	33.9
5	47.4	38.1
6	27.5	27.7
7	34.0	32.7
8	28.4	23.2
9	35.0	35.0
10	20.7	22.7

(i) Explain why these data should be analysed by a paired samples test.

(ii) What underlying distributional assumption is necessary for the paired sample t-test to be appropriate? Carry out this test at the 5% level of significance for the above data. Provide a two-sided 90% confidence interval for the true mean difference between times.

(iii) State the name of, and *briefly* describe, a statistical procedure that could be used in circumstances when the assumption in (ii) is not valid.

19 Steel girders are made by a specialised process which includes a treatment intended to increase the tensile strength of the steel. To check whether this treatment is effective, the tensile strengths of a random sample of seven girders are measured both before and after the treatment. The results (in a suitable unit) are as follows:

Girder	A	B	C	D	E	F	G
Strength before treatment	25.1	23.2	24.7	25.4	24.9	24.0	24.4
Strength after treatment	28.8	25.9	24.5	28.0	24.3	26.3	26.5

Use an appropriate t-test to examine, at the 1% level of significance, whether there is evidence that the treatment has been effective. State carefully the necessary distributional assumption.

Provide a one-sided 95% confidence interval giving a lower bound for the mean increase in strength due to the treatment. State carefully the interpretation of this interval.

[MEI]

20 A large university has many academic departments, research institutes and so on, each of which has a general office where supplies (stationery, computer disks, etc.) are kept. At a certain time, the values of these supplies are found for a random sample of nine of the general offices; the values (in thousands of pounds) are:

6.76 8.19 4.28 5.64 7.10 5.88 3.15 7.20 6.52.

Subsequently, the university management introduces new procedures in an effort to reduce the amount of these supplies held in the general offices. Afterwards, the values are found for a random sample of 12 general offices; the values (in thousands of pounds) are

5.06	5.24	6.50	3.85	2.80	7.16
4.88	5.04	4.67	4.50	5.44	5.10.

Use an appropriate t-test to examine, at the 5% level of significance, whether there is evidence that the average value of the supplies held in the offices has been reduced. State carefully the assumptions on which this analysis is based. Provide a two-sided 95% confidence interval for the true difference in mean overall value before and after the introduction of the new procedures. State carefully the interpretation of this interval.

[MEI]

21 A consumer research organisation is comparing prices of food in two large supermarkets. The prices (in pence) of a random sample of items of food sold in both supermarkets are found to be as follows.

Item	1	2	3	4	5
Store A	232	79	188	56	49
Store B	216	74	208	52	42

Item	6	7	8	9	10
Store A	46	19	37	33	19
Store B	49	18	31	38	17

(i) Use a paired t-test to examine, at the 5% level of significance, the null hypothesis that the mean difference in prices in the two supermarkets is zero.

(ii) Provide a two-sided 90% confidence interval for the mean difference in prices.

(iii) State the distributional assumption underlying your analysis in **(i)** and **(ii)**.

(iv) Explain why it is appropriate for the research organisation to design the investigation in such a way that a paired, rather than an unpaired, procedure is used for analysis.

[MEI]

22 A fermentation process causes the growth of an enzyme. The amount of enzyme present in the mixture after a certain number of hours needs to be measured accurately. An inspector is comparing two procedures for doing this, there being a suspicion that the procedures are leading to different results. Eight samples are therefore taken and each is divided into two sub-samples of which one is randomly assigned for analysis by the first procedure and the other by the second. The data (in a convenient unit of concentration) are as follows.

Sample	Result from first procedure	Result from second procedure
1	214.6	211.8
2	226.2	224.7
3	219.6	219.8
4	208.4	205.2
5	215.1	212.6
6	220.8	218.0
7	218.4	219.2
8	212.3	209.7

It is understood that the underlying populations are satisfactorily modelled by normal distributions.

(i) Use an appropriate t-test to examine these data, stating clearly the null and alternative hypotheses you are testing. Use a 1% significance level.

(ii) Provide a two-sided 95% confidence interval for the difference between the mean results of the procedures.

(iii) Explain why it is sensible to carry out the investigation by dividing samples into two sub-samples as described above, and why the sub-sample to be analysed by the first method should be assigned at random.

[MEI]

1 You should now know how to test the hypothesis that the mean value of some random variable differs in two conditions, in a number of contexts. The test statistics used in the different contexts are given in the following table.

Variances	Sample size	Underlying distribution	Paired test	Unpaired test: variances equal	Unpaired test: variances unequal
Known	Any	Normal	$\dfrac{\overline{D}}{\frac{\sigma}{\sqrt{n}}} \sim N(0,1)$	$\dfrac{\overline{X}-\overline{Y}}{\sigma\sqrt{\frac{1}{n_X}+\frac{1}{n_Y}}} \sim N(0,1)$	$\dfrac{\overline{X}-\overline{Y}}{\sqrt{\frac{\sigma_X^2}{n_X}+\frac{\sigma_Y^2}{n_Y}}} \sim N(0,1)$
Known	Large	Need not be normal	$\dfrac{\overline{D}}{\frac{\sigma}{\sqrt{n}}} \approx N(0,1)$	$\dfrac{\overline{X}-\overline{Y}}{\sigma\sqrt{\frac{1}{n_X}+\frac{1}{n_Y}}} \approx N(0,1)$	$\dfrac{\overline{X}-\overline{Y}}{\sqrt{\frac{\sigma_X^2}{n_X}+\frac{\sigma_Y^2}{n_Y}}} \approx N(0,1)$
Not known	Any	Normal	$\dfrac{\overline{D}}{\frac{S}{\sqrt{n}}} \sim t_{n-1}$	$\dfrac{\overline{X}-\overline{Y}}{S\sqrt{\frac{1}{n_X}+\frac{1}{n_Y}}} \sim t_{n_X+n_Y-2}$	No test discussed here is appropriate.
Not known	Large	Need not be normal	$\dfrac{\overline{D}}{\frac{S}{\sqrt{n}}} \approx N(0,1)$	$\dfrac{\overline{X}-\overline{Y}}{S\sqrt{\frac{1}{n_X}+\frac{1}{n_Y}}} \approx N(0,1)$	$\dfrac{\overline{X}-\overline{Y}}{\sqrt{\frac{S_X^2}{n_X}+\frac{S_Y^2}{n_Y}}} \approx N(0,1)$
Known or not	Small	Need not be normal	No test discussed here is appropriate.		

2 To test the hypothesis that two random variables differ by the amount δ simply replace $\overline{D}$ by $(\overline{D}-\delta)$ or $(\overline{X}-\overline{Y})$ by $(\overline{X}-\overline{Y}-\delta)$ in the numerator of the appropriate statistic.

3 Confidence intervals can be constructed from the test statistic by observing that a confidence interval, in the appropriate situation, for the difference of the means in the two conditions, is given by the interval:
(numerator of statistic) $\pm$ (critical value) $\times$ (denominator of statistic)

4 Note carefully the sections of this table with no appropriate test. It is just as important to know when not to conduct a test as how to do so! Some of these blanks will be filled in Chapter 6, when you look at the Wilcoxon tests.

Non-parametric tests of location

Whenever a large sample of chaotic elements are taken in hand and marshalled in the order of their magnitude, an unsuspected and most beautiful form of regularity proves to have been latent all along.

Sir Francis Galton

Gandhi's popularity was measured by the size of the crowds he drew. These days it is more common for opinion pollsters to ask us questions like the one below.

'The Prime Minister is doing the best possible job, in the circumstances.'

Please choose one of the following responses:

agree strongly	agree	inclined to agree	have no opinion	inclined to disagree	disagree	disagree strongly

You will probably recognise this as the sort of question that is asked by opinion pollsters. However, surveys of people's attitudes are not just undertaken on political issues: market researchers for businesses, local authorities, psychologists and pressure groups, for instance, are all interested in what we think about a very wide variety of issues.

You are going to test the hypothesis:

The Prime Minister's performance is generally disapproved of.

The question above was asked of a group of twelve 17-year-olds and each reply recorded as a number from 1 to 7, where 1 indicates 'agree strongly', and 7 'disagree strongly'. This method of recording responses gives a *rating scale* of attitudes to the Prime Minister's performance. The data obtained are shown below.

$$3 \quad 6 \quad 7 \quad 4 \quad 3 \quad 4 \quad 7 \quad 3 \quad 5 \quad 6 \quad 5 \quad 6$$

What do these data indicate about the validity of the hypothesis in the population from which the sample was drawn?

You should recognise this question as similar to those you asked when conducting *t*-tests. You want to know whether attitudes in the population as a whole are centred around the neutral response of '4' or show lower approval in general. That is, you want a *test of location* of the sample: one which decides what values are taken, on average, in the population. You could not use a *t*-test here to decide whether the mean of the underlying distribution equals 4 because the response variable is clearly not normally distributed: it only takes discrete values from 1 to 7 (and the sample size is small). This chapter looks at some tests of location which are valid even for small samples without the strict distributional assumptions required by the *t*-test and are therefore more widely applicable.

The sign test

One very simple way of handling the sort of data you have here is to make the hypotheses:

H_0: People are equally likely to agree or disagree with the statement.

H_1: People are more likely to disagree than agree.

An opinion has been expressed by ten people (the two whose response is coded as '4' expressed no opinion) and we shall assume that they constitute a random sample from the population. If the null hypothesis is true then each of these people will, independently, agree or disagree with the statement with probability $\frac{1}{2}$. The number agreeing, A, will therefore have a binomial distribution $B(10, \frac{1}{2})$.

To test the (one-tailed) hypothesis at the 5% level you are looking for the greatest value of a that makes $P(A \leqslant a) \leqslant 0.05$. From the tables in the Students' Handbook, using $n = 10$, this critical value is $a = 1$ and so the critical region for the test is $\{0, 1\}$.

In your data, three of those responding agree with the statement, so you accept the null hypothesis that people are equally likely to agree or disagree with the statement.

This test has the advantage of great simplicity, is useful in many circumstances and can be decisive. Because the calculation required is so quick to carry out, it is often useful in an initial exploration of a set of data, and may indeed be all that is necessary.

However, there is information in the sample which is ignored in the sign test: nobody in the sample *strongly* agreed with the statement while, of the seven who disagreed, three did so strongly. The test described in the next section, while it is slightly more complicated, takes this extra information about the attitudes of the sample into account.

The Wilcoxon single sample test

Suppose that, despite your hypothesis, there is no tendency to approve or disapprove of the Prime Minister. Then it would seem plausible that in the population from which the data are drawn the response variable should be modelled as follows.

- It has a median value of 4 – that is, half approve and half disapprove.

- It is symmetrically distributed about this median value – so that, for instance you do not have half strongly approving and half slightly disapproving.

The strategy you adopt, therefore, is to test, on the assumption that responses are symmetrically distributed about the median response, the hypotheses

H_0: The median response is 4.

H_1: The median response is greater than 4.

Note the form of the alternative hypothesis which reflects the fact that our original question was one-tailed ('is the Prime Minister generally disapproved of?').

The Wilcoxon test, like the Spearman's correlation test (see *Statistics 2*), is based on ranks. In the Wilcoxon case, however, you do not rank the actual data themselves, but their distances from the hypothesised median of the population: here, the rating of 4.

For this data this gives the following results.

Rating, r	$r - 4$	$|r - 4|$	Rank
3	−1	1	3
6	2	2	7
7	3	3	9.5
3	−1	1	3
7	3	3	9.5
3	−1	1	3
5	1	1	3
6	2	2	7
5	1	1	3
6	3	2	7

Note

1 Those who gave a rating of 4 are omitted (they expressed no opinion).

2 The ratings are ranked according to their absolute – not their signed – differences from the median.

3 The rating with the smallest difference is given the ranking of 1 and the rating with the largest difference has ranking 10.

4 Where two or more ratings have the same difference, they are given the appropriate 'average rank'. For instance, the five ratings with differences of 1 should occupy the ranks 1, 2, 3, 4, 5 so they are all given the average rank of 3 and the two ratings with differences of 3 should occupy the ranks 9 and 10 so they are both given the average rank of 9.5.

Suppose that the assumption of symmetry is correct, and the null hypothesis that the median is 4 is true. Then you would expect a rating of 5 to come up as often as a rating of 3, a rating of 6 to come up as often as a rating of 2 and a rating of 7 to come up as often as a rating of 1. In other words, ratings at each distance from the supposed median of 4 should be equally likely to be above or below that median.

To test whether the data support this, the next step is to calculate the sum of the ranks of the ratings above 4 and below 4 and compare these with the total sum of the ranks. Here:

sum of ranks of ratings above $4 = 7 + 9.5 + 9.5 + 3 + 7 + 3 + 7 = 46$

sums of ranks of ratings below $4 = 3 + 3 + 3 = 9$

total sums of ranks $= 9 + 46 = 55$.

Note that because of the way in which the ratings were ranked, the total sum of ranks must be equal to the sum of the numbers from 1 to 10:

$$= 1 + 2 + \ldots + 10 = \frac{1}{2}10 \times (10 + 1) = 55.$$

If the null hypothesis were true, you would expect the sum of the ranks of ratings above 4 to be approximately equal to the sum of the ranks of ratings below 4. This means that each would be about half of 55, i.e. 27.5. The fact that for this data the sum of ranks of ratings above 4 is considerably more than this, and the sum of ranks of ratings below 4 correspondingly less, implies either that more people disapproved than approved or that those who disapproved tended to disapprove more, so that the larger rank sum is associated with disapproval. Actually, both of these are true of the data.

In order to conduct an hypothesis test, you need to know the critical values of the test statistic. Tables of the critical values for the Wilcoxon test are available, and a section of one is shown in Figure 6.1. Later in the chapter you will see how these tables are calculated.

1-tail	5%	$2\frac{1}{2}\%$	1%	$\frac{1}{2}\%$
2-tail	10%	5%	2%	1%
n				
2	–	–	–	–
3	–	–	–	–
4	–	–	–	–
5	0	–	–	–
6	2	0	–	–
7	3	2	0	–
8	5	3	1	0
9	8	5	3	1
10	10	8	5	3
11	13	10	7	5
12	17	13	9	7
13	21	17	12	9
14	25	21	15	12
15	30	25	19	15

Figure 6.1 Critical values for the Wilcoxon single sample and paired sample tests

The test you are conducting is one-tailed because you are trying to decide whether your data indicate disapproval; that is, whether the sum of ranks corresponding to approval is significantly smaller and, equivalently, the sum of ranks corresponding to disapproval significantly larger, than chance would suggest. The table is constructed to give the largest value of the rank sum that can be regarded as significantly smaller than chance would suggest, so it is the sum of ranks corresponding to approval (those ratings below 4) that you should compare with the critical value.

Once the two subjects with no opinion are excluded, you have a sample size of 10, and for a one-tailed test at the 5% significance level the table gives a critical value of 10. This means that any value less than or equal to 10 for the sum of the ranks of the ratings below 4 lies in the critical region. Our data give a value of 9 for the sum, so you can reject the null hypothesis in favour of the alternative that the median is greater than 4.

Formal procedure for the Wilcoxon single sample test

To test the hypotheses:

H_0: That the population median of a random variable is equal to a given value M.

H_1: **(a)** That the population median $\neq M$

or **(b)** That the population median $> M$

or **(c)** That the population median $< M$.

Here **(a)** is a two-tailed test, and **(b)** and **(c)** are both one-tailed tests.

Assumption: That the random variable is symmetrically distributed about its median.

Data: The values $x_1, x_2, \ldots, x_n$ of the random variable in a sample of size n. If any of these is equal to M, remove it from the list and reduce n accordingly.

CALCULATION OF THE TEST STATISTIC

1 Calculate the absolute differences between each sample value and the hypothesised median M, i.e.

$$|x_1 - M|, |x_2 - M|, \ldots, |x_n - M|.$$

2 Rank these values from 1 to n, giving the lowest rank to the smallest absolute difference. If a set of absolute difference is equal, each is given the rank which is the average of the ranking positions they occupy together.

3 Calculate the sum W_+ of the ranks of the sample values which are greater than M, and the sum W_- of the ranks of the sample values which are less than M.

4 Check that $W_+ + W_- = \frac{1}{2}n(n+1)$; this must work because the right-hand side is the formula for the sum of the numbers from 1 to n, i.e. the total of all the ranks.

5 The test statistic W is then found as follows:

- for the two-tailed alternative hypothesis **(a)**, take the test statistic W to be the smaller of W_- and W_+

- for the one-tailed alternative hypothesis **(b)**, take $W = W_-$

- for the one-tailed alternative hypothesis **(c)**, take $W = W_+$.

SIGNIFICANCE OF THE TEST STATISTIC

The null hypothesis is rejected if W is *less than or equal to* the appropriate critical value found in the tables which depends on the sample size, the chosen significance level and whether the test is one- or two-tailed.

Note

If the assumption of symmetry of the distribution about its median is true then this median is also the mean.

Rationale for the Wilcoxon test

As noted above, the sum of W_+ and W_- is determined by the sample size, so that the criterion for rejecting the null hypothesis is that the difference between W_+ and W_- is large enough. What will make this difference large? This requires W_+ to contain *more* ranks, or *larger* ranks than W_- (or vice versa). This will occur if *most* of the sample or the *more extreme* values (those furthest from the hypothesised median) in the sample are above the median rather than below (or vice versa). The largest difference between W_+ and W_- will therefore occur if the sample contains a few values just below the hypothesised median and many values well above the hypothesised median (or vice versa). But this is exactly the situation which would cast most doubt on the claim that the suggested median is the true one.

You can also see from this argument why the assumption of symmetry is important. The random variable X, for example, with the highly skewed distribution:

$$P(0) = 0.5, P(2) = P(4) = P(6) = P(8) = P(10) = 0.1$$

has median 1, but samples from this distribution will usually have about half their values just below the median (at 0), but the other half ranging over values substantially above the median (from 2 to 10). Thus large values of W_+ and correspondingly smaller values of W_- will be expected, and should not be taken to cast doubt on the null hypothesis median $= 1$.

EXAMPLE 6.1 A railway Customer Service Division knows from long experience that if passengers are asked to rate a railway company's buffet car service on a scale from 1 to 10 their responses are symmetrically distributed about a median of 4.5.

After an experimental trolley service is introduced on a particular route, passengers are asked to rate this service on a scale from 1 to 10. The ratings of a sample of 16 passengers were as follows.

2	4	1	4	9	3	3	5
6	2	1	2	5	6	2	4

Is there evidence at the 5% level that passengers rate this service differently?

SOLUTION

You are conducting a test of the following hypotheses:

H_0: The ratings are symmetrically distributed about a median of 4.5.

H_1: The ratings are symmetrically distributed about a llmedian different from 4.5.

The first step in calculating W_+ and W_- is to draw up a table giving the absolute differences of the ratings from the hypothesised median.

Rating, r	Frequency	$r - 4.5$	$|r - 4.5|$
1	2	−3.5	3.5
2	4	−2.5	2.5
3	2	−1.5	1.5
4	3	−0.5	0.5
5	2	0.5	0.5
6	2	1.5	1.5
9	1	4.5	4.5

The ranks are calculated from these data as follows.

The smallest absolute difference that you are ranking is 0.5, and this occurs with frequency 5, so the ranks, 1, 2, 3, 4 and 5 are associated with 0.5. The average of these ranks is 3.

The next absolute difference being ranked is 1.5, which occurs with frequency 4, so the ranks 6, 7, 8 and 9 are associated with 1.5. The average of these ranks is 7.5.

A complete list of ranks can now be drawn up by the same method and added to the table.

Rating, r	Frequency	$r - 4.5$	$|r - 4.5|$	Rank
1	2	−3.5	3.5	14.5
2	4	−2.5	2.5	11.5
3	2	−1.5	1.5	7.5
4	3	−0.5	0.5	3
5	2	0.5	0.5	3
6	2	1.5	1.5	7.5
9	1	4.5	4.5	16

The sum of the ranks for the ratings below the median of 4.5 is:

$$W_- = 14.5 \times 2 + 11.5 \times 4 + 7.5 \times 2 + 3 \times 3 = 99$$

and the sum of the ranks for the ratings above the median is:

$$W_+ = 3 \times 2 + 7.5 \times 2 + 16 \times 1 = 37.$$

Check: $W_+ + W_- = 99 + 37 = 136 = \dfrac{1}{2} \times 16 \times 17.$

This is a two-tailed test, so the test statistic, W, is taken to be the smaller of W_+ and W_- which in this case is $W_+ = 37$.

From the tables, the critical value for a two-tailed test on a sample of size 16, using the 5% significance level, is 29. But $37 > 29$ so there is no significant evidence that the rankings of the trolley service are different from those of the buffet car and you accept the null hypothesis.

Why Wilcoxon?

Both the Wilcoxon test and the *t*-test are testing whether the distribution of a random variable in a population has a given value of a *location parameter*. A location parameter is any parameter which, when it varies, shifts the range of values taken by the random variable but not the shape of the distribution. For instance, in the family of normal distributions, the mean is a location parameter but the variance is not; in the family of rectangular distributions, the mid-range is a location parameter but the range is not.

The value of the Wilcoxon test is that it does not make the rather strict distributional assumption of the *t*-test – that the distribution of the random variable is normal. It is therefore very useful when this assumption is not thought to be justified, and when the sample size is not large enough for the sample means nevertheless to be normally distributed. In fact, although the Wilcoxon test places a less severe restriction than the *t*-test on the family of distributions which the underlying variable might possess, it is nonetheless of comparable power to the *t*-test under a wide range of conditions. This means that it is a sensible choice for testing location, even when a *t*-test might also be justifiable.

Calculating critical values

Imagine taking a sample of size six from a distribution which is symmetrical about a value M. Assume for simplicity that no two of the six sample values are equal. A sample value above M can therefore have any of the ranks from 1 to 6. This means that the set of sample values which are above M can correspond to any of the 64 possible subsets of the ranks from 1 to 6. These subsets are listed on page 121, together with the rank sum W_+ which is given by each.

Set	W_+	Set	W_+	Set	W_+	Set	W_+
{}	0	{1}	1	{2}	2	{3}	3
{4}	4	{5}	5	{6}	6	{1,2}	3
{1,3}	4	{1,4}	5	{1,5}	6	{1,6}	7
{2,3}	5	{2,4}	6	{2,5}	7	{2,6}	8
{3,4}	7	{3,5}	8	{3,6}	9	{4,5}	9
{4,6}	10	{5,6}	11	{1,2,3}	6	{1,2,4}	7
{1,2,5}	8	{1,2,6}	9	{1,3,4}	8	{1,3,5}	9
{1,3,6}	10	{1,4,5}	10	{1,4,6}	11	{1,5,6}	12
{2,3,4}	9	{2,3,5}	10	{2,3,6}	11	{2,4,5}	11
{2,4,6}	12	{2,5,6}	13	{3,4,5}	12	{3,4,6}	13
{3,5,6}	14	{4,5,6}	15	{1,2,3,4}	10	{1,2,3,5}	11
{1,2,3,6}	12	{1,2,4,5}	12	{1,2,4,6}	13	{1,2,5,6}	14
{1,3,4,5}	13	{1,3,4,6}	14	{1,3,5,6}	15	{1,4,5,6}	16
{2,3,4,5}	14	{2,3,4,6}	15	{2,3,5,6}	16	{2,4,5,6}	17
{3,4,5,6}	18	{1,2,3,4,5}	15	{1,2,3,4,6}	16	{1,2,3,5,6}	17
{1,2,4,5,6}	18	{1,3,4,5,6}	19	{2,3,4,5,6}	20	{1,2,3,4,5,6}	21

What are the probabilities of each possible sum W_+? The symmetry of the distribution implies that each possible rank from 1 to 6 is equally likely to arise from a sample value above or below M and so the 64 subsets above are all equally likely.

This means you can calculate, for instance:

$$P(W_+ = 5) = \frac{\text{number of subsets where } W_+ = 5}{\text{total number of subsets}} = \frac{3}{64}$$

The complete distribution of the values of W_+ found in this way is:

Value	0	1	2	3	4	5	6	7	8	9	10	11	12	13	14	15	16	17	18	19	20	21
Probability $\times \frac{1}{64}$	1	1	1	2	2	3	4	4	4	5	5	5	5	4	4	4	3	2	2	1	1	1

The cumulative probabilities can also easily be found from the table: the first few, correct to three decimal places, are:

$$P(W_+ \leqslant 0) = \frac{1}{64} = 0.016$$

$$P(W_+ \leqslant 1) = \frac{1}{64} + \frac{1}{64} = \frac{2}{64} = 0.031$$

$$P(W_+ \leqslant 2) = \frac{1}{64} + \frac{1}{64} + \frac{1}{64} = \frac{3}{64} = 0.047$$

$$P(W_+ \leqslant 3) = \frac{1}{64} + \frac{1}{64} + \frac{1}{64} + \frac{2}{64} = \frac{5}{64} = 0.078$$

These cumulative probabilities enable you to find the critical values you require. For instance, note that $P(W_+ \leqslant 2) \leqslant 0.05$, but that $P(W_+ \leqslant 3) > 0.05$. This means that the one-tailed critical value at the 5% level is 2.

ACTIVITY   Check the critical values given in the tables for $n = 6$.

You may have noticed that not all the work in this section was necessary. You only actually used the information shown below from the first two lines of the complete list of rank sums.

Set	W_+	Set	W_+	Set	W_+	Set	W_+
{}	0	{1}	1	{2}	2	{3}	3
						{1,2}	3

ACTIVITY   Find the critical values for the $n = 7$ case, only writing down the rank sums that are necessary.

Normal approximation

The tables only give critical values where the sample size, n, is at most 50. For larger sample sizes you use the fact that, under the null hypothesis, the test statistic, W, is approximately normally distributed with mean $\dfrac{n(n+1)}{4}$ and variance $\dfrac{n(n+1)(2n+1)}{24}$

EXAMPLE 6.2   Determine the 1% one-tailed (2% two-tailed) critical value for a sample of size 84.

SOLUTION

You want to find the integer w so that:

$$P(W \leqslant w) \leqslant 0.01$$

where W has mean $\dfrac{84 \times 85}{4} = 1785$ and variance $\dfrac{84 \times 85 \times 169}{24} = 50\,277.5$.

Note that W is a discrete variable, so that you must make a continuity correction:

$$P(W \leqslant w) \approx P(\text{Normal approximation to } W < w + 0.5).$$ Thus you require:

$$\Phi\left(\frac{w + 0.5 - 1785}{\sqrt{50\,277.5}}\right) \leqslant 0.01$$

so:

$$w \leqslant 1784.5 + \sqrt{50\,277.5}\,\Phi^{-1}(0.01)$$

$$= 1784.5 - \sqrt{50\,277.5}\,\Phi^{-1}(0.99)$$

$$= 1784.5 - \sqrt{50\,277.5} \times 2.326$$

$$= 1262.95$$

This means that values of W less than or equal to 1262 are in the critical region: 1262 is the critical value.

1 An ancient human settlement site in the Harz mountains has been explored by archaeologists over a long period. They have established by a radio-carbon method that the ages of bones found at the site are approximately uniformly distributed between 3250 and 3100 years. A new potassium-argon method of dating has now been developed and eleven samples of bone randomly selected from finds at the site are dated by this new method. The ages, in years, determined by the new method are as listed below.

3115	3234	3247	3198	3177	3226
3124	3204	3166	3194	3220	

Is there evidence at the 5% level that the potassium-argon method is producing different dates, on average, for bones from the site?

2 A local education authority sets a reasoning test to all eleven-year-olds in the borough. The scores of the whole borough on this test have been symmetrically distributed around a median of 24 out of 40 over many years.

One year a primary school's 33 leavers have the following scores out of 40.

21	11	34	32	19	23	26	35	21	35	40
13	15	28	31	26	21	16	24	22	29	36
38	37	27	22	20	18	32	37	29	28	33

Is there evidence at the 5% level to support the headteacher's claim that her leavers score better on the reasoning test than average?

When must this claim have been made if the hypothesis test is to be valid?

3 An investigator stopped a sample of 72 city workers and checked the time shown by their watches against an accurate timer. The number of minutes fast (+) or slow (−) is recorded for each watch, and the data are shown.

+2	0	+4	0	−1	−7	+1	+2	+2
−1	−3	+2	4	−1	0	+3	+2	+3
+1	−1	+8	+4	−2	−4	+5	+1	−2
−3	+2	0	+2	+4	+2	0	+3	−1
−2	−4	+1	+3	+6	+2	0	−6	0
−1	−2	+1	+2	+2	−3	+2	0	+2
−1	+2	+2	−5	−1	0	+5	+2	+3
+1	−2	+9	+4	−2	−3	+4	+1	−2

Is there evidence at the 5% level that city workers tend to keep their watches running fast?

4 Becotide inhalers for asthmatics are supposed to deliver 50 mg of the active ingredient per puff. In a test in a government laboratory, 17 puffs from randomly selected inhalers in a batch were tested and the amount of active ingredient that was delivered was determined. The results, in mg, are given below.

43	47	52	51	44	50	51	41	48
46	52	50	47	45	49	46	42	

Is there evidence at the 2% level that the inhalers are not delivering the correct amount of active ingredient per puff?

5 When a consignment of grain arrives at Rotterdam docks the percentage of moisture in 11 samples is measured. It is claimed that when the ship left Ontario, the percentage of moisture in the grain was 2.353% on average. The percentages found in the samples were as follows

5.294	0.824	3.353	1.706	3.765	3.235
8.235	0.76	3.412	6.471	3.471	

(i) Test at the 5% level whether the median percentage of moisture in the grain is greater than 2.353, using the Wilcoxon single sample test. What assumptions are you making about the distribution of the percentage of moisture in the grain?

(ii) Test at the 5% level whether the mean percentage of moisture in the grain is greater than 2.353, using the *t*-test. What assumption are you making about the distribution of the percentage of moisture in the grain?

(iii) Compare your two conclusions and comment.

6 Check the claim in the tables that the critical value for the Wilcoxon single sample test, at the 5% level, for a sample of size 9 is 8.

Hint: There are $2^9 = 512$ different sets of ranks that, under the null hypothesis, are equally likely to make up W_+ but it is only necessary to write down those with the smallest rank sums – the sets giving rank sums up to and including 9 are sufficient to verify the result.

7 An estate agent claims that the median price of detached houses in a certain area is £115 000. An advertising standards officer takes a random sample of ten such houses that have recently been sold and finds that their prices (in £) are

| 138 000 | 110 000 | 117 500 | 130 000 | 121 900 |
| 106 000 | 165 000 | 134 000 | 129 500 | 125 000. |

(i) Suggest why a normal distribution might not be a suitable model for the underlying population of prices.

(ii) Calculate the value of the Wilcoxon single sample test statistic for examining the estate agent's claim.

(iii) Taking the alternative hypothesis to be that the median price is not £115 000, use the appropriate table in the Students' Handbook to state the critical region for a test at

  (a) the 5% level of significance

  (b) the 1% level of significance.

(iv) What is your conclusion in respect of the estate agent's claim?

(v) Find the level of significance of the data as given by the normal approximation

$$N\left(\frac{n(n+1)}{4}, \frac{n(n+1)(2n+1)}{24}\right)$$

to the distribution of the test statistic if the null hypothesis is true. Comment on the accuracy of the approximation in the light of your answers to **(iii)**.

[MEI]

The Wilcoxon paired sample test

A few years ago, some people believed that Maths and English GCSE exams were of different levels of difficulty and it was claimed that, in the country as a whole, candidates were getting, on average, at least one-and-a-half grades higher in English than in Maths. A group of eleven students who took their GCSE exams at that time gained the following results in Maths and English:

Student	1	2	3	4	5	6	7	8	9	10	11
Maths grade	A	D	F	F	C	G	U	F	D	B	E
English grade	B	C	C	A	D	E	G	B	C	E	E

Are those claiming a difference of as much as one-and-a-half grades justified?

For each student, you can calculate how many grades better their English result was than their Maths result.

Student	1	2	3	4	5	6	7	8	9	10	11
Grades better in English	−1	1	3	5	−1	2	1	4	1	−3	0

The hypotheses are:

H_0: The median number of grades better in English is 1.5.

H_1: The median number of grades better in English is less than 1.5.

The Wilcoxon one-sample test described above can now be used. The number of grades better in English is the value, g, of the random variable, G, which you are assuming is symmetrically distributed about its hypothesised median.

Student	g	$g - 1.5$	$|g - 1.5|$	Rank
1	−1	−2.5	2.5	8
2	1	−0.5	0.5	2.5
3	3	1.5	1.5	5.5
4	5	3.5	3.5	10
5	−1	−2.5	2.5	8
6	2	0.5	0.5	2.5
7	1	−0.5	0.5	2.5
8	4	2.5	2.5	8
9	1	−0.5	0.5	2.5
10	−3	−4.5	4.5	11
11	0	−1.5	1.5	5.5

So that : $W_+ = 5.5 + 10 + 2.5 + 8 = 26$

$$W_- = 8 + 2.5 + 8 + 2.5 + 2.5 + 11 + 5.5 = 40.$$

Check: $W_+ + W_- = 26 + 40 = 66 = \dfrac{1}{2} \times 11 \times 12$ as required.

The test statistic is $W = W_+ = 26$.

From the tables, if you use the 5% significance level, the critical value for a sample size of 11 is 13. Since $13 < 26$ you accept the null hypothesis: you have no good evidence to suggest that the claim is incorrect.

In general, when the values of a random variable have been measured on a sample in two different conditions, you may want to test the hypothesis that the medians differ by some given amount between the two conditions.

In this situation the Wilcoxon paired sample test procedure is:

1 Find the differences between the values in the two conditions.

2 Use the Wilcoxon single sample test with the hypothesis that these differences have the suggested median.

The distributional assumption is then that, in the population, the differences between the values of the random variables in the two conditions are symmetrically distributed about the median difference.

Perhaps the most natural context in which this test is used is when you are trying to detect a shift in a location parameter of the distribution of a random variable between two conditions.

It is worth noticing that the Wilcoxon paired sample test and the Wilcoxon single sample test are related in the same way as the paired sample *t*-test and the single sample *t*-test.

EXAMPLE 6.3

Seven randomly selected economists were asked on two occasions to predict what the annual growth rate of GDP would be in December 1995. Their predictions, made in June 1994 and December 1994 were as listed below.

Economist	1	2	3	4	5	6	7
June 1994 prediction (%)	2.2	3.7	2.1	2.3	3.4	2.5	2.1
December 1994 prediction (%)	2.8	3.8	2.6	3.1	3.0	2.6	2.7

Is there evidence at the 5% level that economists have become more optimistic between June and December about the future growth rate?

SOLUTION

The increases in the predictions between June and December 1994 are:

Economist	1	2	3	4	5	6	7
Increase in prediction (%)	0.6	0.1	0.5	0.8	−0.4	0.1	0.6

The hypotheses under test are:

H_0: The median increase in prediction is zero.

H_1: The median increase in prediction is positive.

So it is the absolute increases in prediction themselves that you need to rank.

Economist	1	2	3	4	5	6	7
Increase in prediction (%)	0.6	0.1	0.5	0.8	−0.4	0.1	0.6
Absolute increase in prediction (%)	0.6	0.1	0.5	0.8	0.4	0.1	0.6
Rank	5.5	1.5	4	7	3	1.5	5.5

Hence:

$$W_+ = 5.5 + 1.5 + 4 + 7 + 1.5 + 5.5 = 25$$

$$W_- = 3$$

Check: $W_+ + W_- = 25 + 3 = 28 = \dfrac{1}{2} \times 7 \times 8.$

The test statistic is therefore: $W = W_- = 3$

From the tables the critical value at the 5% level for a sample of size 7 is 3 and $3 \leqslant 3$, so that the test statistic lies in the critical region and you reject the null hypothesis in favour of the alternative that economists have become more optimistic.

The Wilcoxon rank sum test

In December 1994, nine American economists made predictions for the growth rate in American GDP for December 1995. Their predictions are shown below, together with seven British predictions for the growth rate in the British GDP made at the same time.

Predictions for December 1995 growth rate

America	3.5	4.2	2.8	3.2	3.7	2.9	3.4	2.8	3.7
Britain	2.8	3.8	2.6	3.1	3.0	2.6	2.7		

Is there evidence that the American economists are more optimistic than their British counterparts about their respective countries' growth rates?

The formal statement of the hypotheses under test is:

H_0: The median difference between American and British economists' predictions is zero.

H_1: The median difference between American and British economists' predictions is positive.

You can get a feeling for these data visually by ranking all 16 predictions, giving something akin to a back-to-back stem-and-leaf diagram.

American predictions	Rank	British predictions
	1.5	2.6, 2.6
	3	2.7
2.8, 2.8	5	2.8
2.9	7	
	8	3.0
	9	3.1
3.2	10	
3.4	11	
3.5	12	
3.7, 3.7	13.5	
	15	3.8
4.2	16	

This makes it look as though the American predictions are higher on the whole: how can you measure how much higher? If you add the ranks of the American and British predictions separately you get the Wilcoxon statistics:

$$W_A = 5 + 5 + 7 + 10 + 11 + 12 + 13.5 + 13.5 + 16 = 93$$

$$W_B = 1.5 + 1.5 + 3 + 5 + 8 + 9 + 15 = 43$$

Check: $W_A + W_B = 93 + 43 = 136 = \dfrac{1}{2} \times 16 \times 17$.

The larger rank sum for the American economists does not, however, merely reflect their larger predictions, but also the fact that they made more predictions. This effect can be removed by subtracting from each rank sum an amount equal to $\dfrac{1}{2}N(N + 1)$, where N is the respective sample size, to obtain adjusted rank totals known as the *Mann–Whitney T values*.

JUSTIFICATION

If there are N_A American economists and N_B British economists, then there are $(N_A + N_B)$ economists altogether and so the total of all the ranks (i.e. all the numbers from 1 to $(N_A + N_B)$) is $\dfrac{1}{2}(N_A + N_B)(N_A + N_B + 1)$.

If the ranking of the economists' predictions were random you would expect the total of the American economists' ranks to be in proportion to their numbers, that is:

$$\frac{N_A}{N_A + N_B} \times \frac{1}{2}(N_A + N_B)(N_A + N_B + 1) = \frac{1}{2}N_A(N_A + N_B + 1)$$

while for the total of the British economists' ranks you would similarly expect:

$$\frac{1}{2}N_B(N_A + N_B + 1)$$

These are not equal when N_A and N_B are not the same, but the subtraction of the suggested terms:

$$\frac{1}{2}N_A(N_A + N_B + 1) - \frac{1}{2}N_A(N_A + 1) = \frac{1}{2}N_A N_B$$

$$\frac{1}{2}N_B(N_A + N_B + 1) - \frac{1}{2}N_B(N_B + 1) = \frac{1}{2}N_A N_B$$

leaves equal 'adjusted totals', as required.

In this case you get the Mann–Whitney T values:

$$T_A = 93 - \frac{1}{2} \times 9 \times 10 = 93 - 45 = 48$$

$$T_B = 43 - \frac{1}{2} \times 7 \times 8 = 43 - 28 = 15$$

You can now see numerically that, on a comparable basis, the British economists did tend to produce predictions of lower rank. So, to complete the hypothesis test, all you need to know is whether the value of 15 is small enough to arise only rarely by chance when the underlying distribution of predictions is identical for Britain and America. An extract from tables giving critical values for the Mann–Whitney statistic is shown in Figure 6.2. As in the one-sample case, the process by which these tables are constructed is demonstrated later in this chapter.

In the tables, m is the size of the smaller sample (here 7) and n the larger (here 9). The Mann–Whitney statistic is the smaller of the two T values (here 15) and its critical value (for a one-tailed test) at the 5% level is also shown as 15. So you reject the null hypothesis that there is no difference between the distributions of predictions in America and Britain in favour of the alternative that the American economists are more optimistic.

The Wilcoxon rank sum test is appropriate for tests of hypotheses similar to those for which the unpaired sample tests of Chapters 3 and 5 were used. Note the absence of the special cases that you had to consider there for small or large samples, equal or unequal variances, normal or non-normal underlying distributions. This *non-parametric* test is applicable in one form to a wide variety of situations where *difference of location* in two conditions is being tested.

Formal procedure for the Wilcoxon rank sum test

Distributional assumption: That the random variables X and Y have distributions with the same shape, though not necessarily identically located.

H_0: That the two random variables also have the same median.

H_1: **(a)** That the median of X is different from that of Y

or **(b)** That the median of X is larger than that of Y

or **(c)** That the median of Y is larger than that of X.

1-tail	5%	$2\frac{1}{2}\%$	1%	$\frac{1}{2}\%$		1-tail	5%	$2\frac{1}{2}\%$	1%	$\frac{1}{2}\%$
2-tail	10%	5%	2%	1%		2-tail	10%	5%	2%	1%
m *n*						*m* *n*				
5 5	4	2	1	0		6 10	14	11	8	6
5 6	5	3	2	1		6 11	16	13	9	7
5 7	6	5	3	1		6 12	17	14	11	9
5 8	8	6	4	2		6 13	19	16	12	10
5 9	9	7	5	3		6 14	21	17	13	11
5 10	11	8	6	4		6 15	23	19	15	12
5 11	12	9	7	5		6 16	25	21	16	13
5 12	13	11	8	6		6 17	26	22	18	15
5 13	15	12	9	7		6 18	28	24	19	16
5 14	16	13	10	7		6 19	30	25	20	17
5 15	18	14	11	8		6 20	32	27	22	18
5 16	19	15	12	9		6 21	34	29	23	19
5 17	20	17	13	10		6 22	36	30	24	21
5 18	22	18	14	11		6 23	37	32	26	22
5 19	23	19	15	12		6 24	39	33	27	23
5 20	25	20	16	13		6 25	41	35	29	24
5 21	26	22	17	14		7 7	11	8	6	4
5 22	28	23	18	14		7 8	13	10	7	6
5 23	29	24	19	15		7 9	15	12	9	7
5 24	30	25	20	16		7 10	17	14	11	9
5 25	32	27	21	17		7 11	19	16	12	10
6 6	7	5	3	2		7 12	21	18	14	12
6 7	8	6	4	3		7 13	24	20	16	13
6 8	10	8	6	4		7 14	26	22	17	15
6 9	12	10	7	5		7 15	28	24	19	16
						7 16	30	26	21	18
						7 17	33	28	23	19
						7 18	35	30	24	21
						7 19	37	32	26	22
						7 20	39	34	28	24
						7 21	41	36	30	25
						7 22	44	38	31	27
						7 23	46	40	33	29
						7 24	48	42	35	30
						7 25	50	44	36	32

Figure 6.2 Critical values for the Mann–Whitney/Wilcoxon rank sum 2-sample test

Note that **(a)** is a two-tailed alternative, while **(b)** and **(c)** are both one-tailed.

Data: The values $x_1, x_2, \ldots, x_m$ of the random variable X in a sample of size m and the values $y_1, y_2, \ldots, y_n$ of the random variable Y in a sample of size n. It is conventional to take $m \leqslant n$ (this affects which random variable is called X and which Y).

CALCULATION OF THE TEST STATISTIC

1 Rank all $(m + n)$ sample values 1 to $(m + n)$, giving the lowest rank to the smallest value. If a set of sample values is equal, each is given the rank which is the average of the ranking positions they occupy together.

2 Calculate the sum of W_X of the ranks of the x values, and the sum of W_Y of the ranks of the y values.

3 Check that $W_X + W_Y = \dfrac{1}{2}(m + n)(m + n + 1)$.

This must work because the right-hand side is the formula for the sum of the numbers from 1 to $(m + n)$, i.e. the total of all the ranks.

4 Calculate the Mann–Whitney T values:

$$T_X = W_X - \frac{1}{2}m(m + 1)$$
$$T_Y = W_Y - \frac{1}{2}n(n + 1).$$

5 Check that $T_X + T_Y = mn$

6 The test statistic T is then found as follows:

- for the two-tailed alternative hypothesis **(a)**, the test statistic T is the smaller of T_X and T_Y

- for the one-tailed alternative hypothesis **(b)**, the test statistic $T = T_Y$

- for the one-tailed alternative hypothesis **(c)**, the test statistic $T = T_X$.

SIGNIFICANCE OF THE TEST STATISTIC

The null hypothesis is then rejected if T is less than or equal to the appropriate critical value. This critical value is found in the tables according to the two sample sizes, the chosen significance level and whether the test is one- or two-tailed.

EXAMPLE 6.4

In an experiment on cultural differences in interpreting facial expressions, 16 English children and 23 Japanese children were each shown 40 full-face photographs of American children who had been asked to display particular emotions in their facial expressions.

They were asked to describe the feelings of the children in the photographs and the number of correct responses they made was noted. The results of this experiment, as scores out of 40, are shown below.

English children	34	27	39	29	33	34
	39	40	27	23	34	29
	37	37	31	35		
Japanese children	32	40	27	25	35	38
	27	38	36	23	28	31
	30	37	27	30	34	40
	26	30	35	32	30	

Is there evidence at the 5% level, that English and Japanese children differ in their abilities to identify American facial expressions?

SOLUTION

The formal statement of the hypotheses under test is:

H_0: The distributions of English and Japanese scores have the same median.

H_1: The distributions of English and Japanese scores have different medians.

under the assumption that the distributions have the same shape.

The scores are shown below, with their ranks.

Japanese scores	Rank	English scores
23	1.5	23
25	3	
26	4	
27, 27, 27	7	27, 27
28	10	
	11.5	29, 29
30, 30, 30, 30	14.5	
31	17.5	31
32, 32	19.5	
	21	33
34	23.5	34, 34, 34
35, 35	27	35
36	29	
37	31	37, 37
38, 38	33.5	
	35.5	39, 39
40, 40	38	40

The rank sums are:

$$W_J = 434.5$$

$$W_E = 345.5$$

Check: $W_J + W_E = 434.5 + 345.5 = 780 = \dfrac{1}{2} \times 39 \times 40.$

So the Mann–Whitney T values are:

$$T_J = 434.5 - \frac{1}{2} \times 23 \times 24 = 434.5 - 276 = 158.5$$

$$T_E = 345.5 - \frac{1}{2} \times 16 \times 17 = 345.5 - 136 = 209.5$$

Since this is a two-tailed test, the test statistic, T, is the smaller of T_J and T_E which is 158.5.

From the tables, the critical value for sample sizes $m = 16$, $n = 23$ at the two-tailed 5% level is 115. Since $115 < 158.5$ you accept the null hypothesis that there is no difference in the distribution of scores between English and Japanese children.

An alternative method for calculating T

There is an alternative way of calculating T, called the Mann–Whitney Method, after the statisticians who devised it independently of Wilcoxon. Both methods are used by statisticians, so both are worth knowing.

The data for a Wilcoxon rank sum test consist of two samples. The smaller sample, of size m, is referred to as sample 1 and the larger, of size n, as sample 2. The Wilcoxon rank sum statistic, T_1, for the smaller sample is calculated by the Mann–Whitney method as follows.

- For each result in sample 1:
    - Count the number of results in sample 2 which are smaller than it.
    - Add to this half the number of results in sample 2 which are equal to it.
- Add up the numbers obtained this way for each result in sample 1 to obtain T_1.

EXAMPLE 6.5

Consider the data in Example 6.4, where $m = 16$ and $n = 23$. The data is reproduced below, in increasing order. Calculate the Mann–Whitney T values.

English children	23	27	27	29	29	31	33	34	34	34	35	37
	37	39	39	40								
Japanese children	23	25	26	27	27	27	28	30	30	30	30	31
	32	32	34	35	35	36	37	38	38	40	40	

SOLUTION

Here:

T_E = the number of Japanese children with scores smaller than 23

(plus half the number with scores equal to 23)

+ the number of Japanese children with scores smaller than 27

(plus half the number with scores equal to 27)

+ ...

+ the number of Japanese children with scores smaller than 40

(plus half the number with scores equal to 40).

These are easy to read off from the ordered lists: for example, there are 14 Japanese children with scores smaller than 34 and one with a score equal to 34. This means that each of the three scores of 34 in the 'English' list contributes $\left(14 + 1 \times \dfrac{1}{2}\right)$ to the value of T_E.

Calculating all the required terms in this way gives:

$$T_E = 0.5 + 4.5 + 4.5 + 7 + 7 + 11.5 + 14 + 14.5 + 14.5 + 14.5 + 16 + 18.5 + 18.5$$

$$+ 21 + 21 + 22$$

$$= 209.5$$

which is the same as the result obtained using the rank sum method.

When using this method, it is usual to obtain the other T value from the fact that $T_J + T_E = mn$, so that:

$$T_J = mn - T_E = 16 \times 23 - 209.5 = 158.5$$

as before.

Critical values for the Wilcoxon rank sum test

In this section you will see how the tables of critical values for this test can be calculated.

Imagine the situation where you have samples of sizes $m = 4$ and $n = 7$. Of the eleven possible ranks from 1 to 11, four must be assigned to the sample of size 4. The smallest values of W_X, the rank sum for the smaller sample, that can arise, and the rank sums which give rise to these, are shown on page 135.

10	{1,2,3,4}				
11	{1,2,3,5}				
12	{1,2,3,6}	{1,2,4,5}			
13	{1,2,3,7}	{1,2,4,6}	{1,3,4,5}		
14	{1,2,3,8}	{1,2,4,7}	{1,2,5,6}	{1,3,4,6}	{2,3,4,5}
15	{1,2,3,9}	{1,2,4,8}	{1,2,5,7}	{1,3,4,7}	{1,3,5,6}	{1,3,4,6}

Altogether, there are $\binom{11}{4} = 330$ ways of picking four of the ranks to be in the W_X rank sum, and under the null hypothesis the ranks from 1 to 11 are assigned randomly to the two groups so that each of these possible assignments is equally likely. This means that the probability of each possible rank sum is the number of different ways of making that sum, as a fraction of 330. Thus you can calculate the cumulative probabilities of the smallest possible values of W_X as:

$$P(\text{rank sum} = 10) = \frac{1}{330} \qquad\qquad = 0.003\,03$$

$$P(\text{rank sum} \leqslant 11) = \frac{1+1}{330} = \frac{2}{330} \qquad\qquad = 0.006\,06$$

$$P(\text{rank sum} \leqslant 12) = \frac{1+1+2}{330} = \frac{4}{330} \qquad\qquad = 0.012\,12$$

$$P(\text{rank sum} \leqslant 13) = \frac{1+1+2+3}{330} = \frac{7}{330} \qquad\qquad = 0.021\,21$$

$$P(\text{rank sum} \leqslant 14) = \frac{1+1+2+3+5}{330} = \frac{12}{330} \qquad\qquad = 0.036\,36$$

$$P(\text{rank sum} \leqslant 15) = \frac{1+1+2+3+5+6}{330} = \frac{18}{330} = 0.054\,55.$$

The critical value of W_X at the 5% level is therefore 14, so the critical value of T is $14 - \frac{1}{2} \times 4 \times 5 = 4$.

For the W_Y, the rank sum for the larger sample, the smallest values and their corresponding rank sums are:

28	{1,2,3,4,5,6,7}				
29	{1,2,3,4,5,6,8}				
30	{1,2,3,4,5,6,9}	{1,2,3,4,5,7,8}			
31	{1,2,3,4,5,6,10}	{1,2,3,4,5,7,9}	{1,2,3,4,6,7,8}		
32	{1,2,3,4,5,6,11}	{1,2,3,4,5,7,10}	{1,2,3,4,5,8,9}	{1,2,3,4,6,7,9}	{1,2,3,5,6,7,8}
33	{1,2,3,4,5,6,12}	{1,2,3,4,5,7,11}	{1,2,3,4,5,8,10}	{1,2,3,4,6,7,10}	{1,2,3,4,6,8,9}
	{1,2,3,5,6,7,9}				

Because the numbers of sets of ranks giving each of the smallest possible totals are the same as for W_X, the critical value for W_Y at the 5% level will also be the fifth smallest of the possible sums: in this case, 32. This gives the critical value of T as $32 - \frac{1}{2} \times 7 \times 8 = 4$ which is the same as before – as it must be since the same critical value is used whether T_X or T_Y is the smaller.

Normal approximation

The tables only give critical values for sample sizes up to 25. For larger sample sizes you use the fact that, under the null hypothesis, the test statistic T is approximately normally distributed with mean $\dfrac{mn}{2}$ and variance $\dfrac{mn(m+n+1)}{12}$.

EXAMPLE 6.6

Determine the 1% one-tailed (2% two-tailed) critical value for the samples of sizes $m = 33$, $n = 58$.

SOLUTION

You want to find the integer t so that:

$$P(T \leqslant t) \leqslant 0.01$$

where T has mean $\dfrac{33 \times 58}{2} = 957$ and variance $\dfrac{33 \times 58 \times (33 + 58 + 1)}{12} = 14\,674$.

Note that T is a discrete variable, so that you must make a continuity correction:

$$P(T \leqslant t) \approx P(\text{Normal approximation to } T < t + 0.5).$$

Thus you require:

$$\Phi\left(\frac{t + 0.5 - 957}{\sqrt{14\,674}}\right) \leqslant 0.01$$

so

$$t \leqslant 956.5 + \sqrt{14\,674}\,\Phi^{-1}(0.01)$$

$$= 956.5 - \sqrt{14\,674}\,\Phi^{-1}(0.99)$$

$$= 956.5 - \sqrt{14\,674} \times 2.326$$

$$= 674.74.$$

This means that values of T less than or equal to 674 are in the critical region: 674 is the critical value.

1 The amount of lead in airborne dust (in parts per million) at 23 sampling spots around London was measured before and after government measures to encourage the use of lead-free petrol were introduced. Does the data below give evidence at the 2.5% level that the government measures have reduced the average amount of lead in the air?

Amount of lead

Before	After
43	47
11	8
133	102
57	51
28	34
91	72
48	41
90	99
205	196
37	37
81	15
111	104
23	56
29	17
78	59
170	138
53	62
61	56
14	13
40	27
167	158
80	97
19	8

2 The speed with which 13 subjects react to a stimulus is timed in hundredths of a second. They then play a computer game for 20 minutes and their reaction times are re-measured. Is there evidence at the 5% level that playing the computer game has reduced their reaction times?

Subject	Reaction time	
	before game	after game
A	231	201
B	337	346
C	168	183
D	243	215
E	197	188
F	205	181
G	265	291
H	170	175
I	302	281
J	250	242
K	316	306
L	252	211
M	226	198

3 Of the 32 leavers from a school who went on to university, 11 went to Oxford or Cambridge and 21 to other universities. Three years later, when these 32 got jobs, their starting salaries, in £, were as listed below.

Oxford and Cambridge students

9012	10760	11405	11617	12030
13040	13772	14200	16430	17500
21650				

Students from other universities

7860	9540	10300	10340	10980
11040	11650	11790	11995	12000
12545	12800	12950	13270	13280
14350	14608	14980	15200	15330
18750				

(i) Test, at the 5% level, the hypothesis that the starting salaries of Oxford and Cambridge graduates are higher than those of graduates from other universities.

(ii) Criticise the sampling procedure adopted in collecting these data.

4 The blood cholesterol levels of 46 men and 28 women are measured. These data are shown below.

Men				
	621	237	92	745
	301	550	182	723
	1301	56	104	428
	209	478	119	303
	417	869	384	1058
	939	1080	1104	829
	1061	145	382	919
	813	770	312	205
	610	139	206	174
	67	258	333	1203
	407	826	810	922
	717	106		
Women	208	529	104	72
	377	482	50	620
	1003	162	94	391
	149	371	208	194
	901	205	370	871
	710	973	304	189
	683	191	233	127

Is there evidence at the 5% level that the blood cholesterol levels of men and women differ? Do the data suggest that the assumptions of the test are justified?

5 Given two rock samples A and B, geologists say that A is harder than B if, when the two samples are rubbed together, A scratches B. Eleven rock samples from stratum X and seven from stratum Y are tested in this way to determine the order of hardness. The list below shows 18 samples in order of hardness and which stratum they came from. The hardest is on the right.

X X X X X Y X Y X X Y X Y Y X X Y Y

Is there evidence at the 5% level that the rock in Stratum Y is harder than that in stratum X?

6 Check the claim in the tables that the critical value for the Wilcoxon rank sum test, at the 2.5% level, for samples of sizes $m = 5$, $n = 8$, is 6.

7 An aggregate material used for road surfacing contains some small stones of various types. A highway engineer is examining the composition of this material as delivered from two separate suppliers. The stones are broadly classified into two types, rounded and non-rounded, and the percentage of the rounded type is found in each of several samples. It is desired to examine whether the two suppliers are similar in respect of this percentage.

(i) In an initial investigation, histograms are drawn of the proportions of rounded stones in samples from the two suppliers. Discuss briefly what these histograms should indicate about the shape of the underlying distribution so that the comparison may reasonably be made using

(a) a t-test

(b) a Wilcoxon rank sum test.

(ii) It is decided that a Wilcoxon rank sum test must be used. Detailed data of the percentage of the rounded type of stone for 15 samples, 9 from one supplier and 6 from the other, are as follows:

Supplier 1	46	52	34	17	21	63	55	48	25
Supplier 2	59	53	71	39	66	58			

Test at the 5% level of significance whether it is reasonable to assume that the true median percentages for the suppliers are the same.

8 Random samples $x_1, x_2, \ldots, x_m$ and $y_1, y_2, \ldots, y_n$ are taken from two independent populations. The Wilcoxon rank sum test is to be used to test the hypotheses:

H_0: The two populations have identical distributions.

H_1: The two populations have identical distributions except that their location parameters differ.

Accordingly, the complete set of $m + n$ observations is ranked in ascending order (it may be assumed that no two observations are exactly equal). S denotes the sum of the ranks corresponding to $x_1, x_2, \ldots, x_m$.

(i) Consider the case $m = 4$, $n = 5$.

   (a) Show that, if all the xs are less than all the ys, the value of S is 10.

   (b) List all possible sets of ranks of the xs that give rise to a value of S such that $S \leqslant 12$.

   (c) What is the total number of ways of assigning four ranks from the available nine to the xs?

   (d) Deduce from your answers to **(b)** and **(c)** that the probability that $S \leqslant 12$ if H_0 is true is $\frac{4}{126}$.

   (e) Compute the value of this probability as given by the normal approximation

$$N\left(\tfrac{1}{2}m(m + n + 1), \tfrac{1}{12}mn(m + n + 1)\right)$$

   to the distribution of S if H_0 is true.

(ii) The following are the numerical values of the data for a case with $m = 6$, $n = 8$:

Sample 1	4.6	6.6	6.0	5.2
$(x_1, x_2, \ldots, x_m)$	8.1	9.5		
Sample 2	5.5	7.9	7.1	6.3
$(y_1, y_2, \ldots, y_n)$	8.4	6.8	10.2	9.0

Test H_0 against H_1 at the 5% level of significance, using the normal approximation given in part **(e)** above to the distribution of S under H_0 or otherwise.

9 As part of the procedure for interviewing job applicants, a firm uses aptitude tests. Each applicant takes a test and receives a score. Two different tests, A and B, are used. The distributions of scores for each test, over the whole population of applicants, are understood to be similar in shape, approximately symmetrical, but not normal. However, the location parameters of these distributions may differ. The personnel manager is investigating this by considering the medians of the distributions, with the null hypothesis H_0 that these medians are equal. It is thought that test B might lead to consistently lower scores, so the alternative hypothesis H_1 is that the median for test B is *less than* that for test A.

The scores from test A for a random sample of seven applicants for a particular job are as follows:

37.6   34.3   38.5   38.8   35.8   38.0   38.2.

The scores from test B for a separate random sample of seven applicants for this job are as follows:

37.3   33.0   33.9   32.1   37.0   35.0   36.2.

(i) Calculate the value of the Wilcoxon rank sum test statistic for these data.

(ii) State the critical point for a one-sided 5% test of H_0 against H_1.

(iii) State whether values *less than* this critical point would lead to the acceptance or rejection of H_0.

(iv) Carry out the test.

Now suppose, instead, that the two distributions can be taken as normal (with the same variance), and that the test is to be conducted in terms of the means. Describe briefly a procedure based on the t-distribution for carrying out this test using these data.

10 Random samples $x_1, x_2, \ldots, x_m$ and $y_1, y_2, \ldots, y_n$ are taken from two independent populations. It is understood that these two populations have identical distributions except possibly for a difference in location parameter. The null hypothesis H_0 that they have the same location parameter is to be tested against the alternative hypothesis H_1 that their location parameters differ, using the Wilcoxon rank sum test. The complete set of all $m + n$ observations is ranked in ascending order (it may be assumed that no two observations are exactly equal); W denotes the sum of the ranks corresponding to $x_1, x_2, \ldots, x_m$.

(i) Show that the minimum value W can take is $\frac{1}{2}m(m + 1)$.

(ii) Find the maximum value W can take.

(iii) The sample data are as follows:

Sample A	10.2	13.5	15.2	15.4	10.8
$(x_1, x_2, \ldots, x_m)$	17.7	12.1	13.8		
Sample B	14.6	19.3	11.6	12.9	18.4
$(y_1, y_2, \ldots, y_n)$	16.9	18.2	18.7	16.1	19.0

Calculate the value of W and hence carry out a two-sided 5% test.

(iv) Show that if H_0 is true the expected value of W is $\frac{1}{2}m(m + n + 1)$. Given that the variance of W if H_0 is true is $\frac{1}{12}m(m + 1)$, repeat the test in **(iii)** using a normal approximation to the distribution of W under H_0.

11 A paired comparison situation is being investigated in which underlying normality cannot be assumed. The Wilcoxon paired sample test is therefore to be used to test the null hypothesis H_0 that the location parameters of the two underlying populations are equal against the alternative hypothesis H_1 that they are not equal.

The observations in the sample from one population are denoted by $x_1, x_2, \ldots, x_m$. The corresponding observations in the other sample are denoted by $y_1, y_2, \ldots, y_n$. The respective differences are denoted by z_i where $z_i = x_i - y_i$ for $i = 1, 2, \ldots, n$. The absolute values of the z_i are ranked (it may be assumed that no two $|z_i|$ are exactly equal). The quantity T is the sum of these ranks for those z_i that are positive.

(i) Find the value of T and carry out the test, at the 5% level of significance, for the following data:

x_i	6.8	7.2	5.5	9.4
	8.8	8.2	7.7	6.6
y_i	8.4	8.5	6.9	9.1
	9.6	7.6	9.7	8.5

(ii) Show that, if H_0 is true, the expected value of T is $\frac{1}{4}n(n + 1)$.

(iii) Find the *actual* level of significance for the data in **(i)** as given by the normal approximation

$$N\left(\tfrac{1}{4}n(n + 1), \tfrac{1}{24}n(n + 1)(2n + 1)\right)$$

to the distribution of T under H_0.

12 A highly skilled typist is comparing the abilities of two word-processing systems for the typing of complicated mathematical expressions. Although both systems provide special facilities, they both have some difficulties in dealing with such expressions. The typist has taken several mathematical articles that have been professionally typeset and has typed each one using both word-processors. For each article, she has counted the number of such expressions with which, in her opinion, there has been unusual difficulty. The results are as follows:

Article	A	B	C	D	E
Word-processor I	2	15	21	6	1
Word-processor II	1	20	29	2	11

Article	F	G	H	I	J
Word-processor I	3	10	14	7	7
Word-processor II	5	3	25	20	22

Underlying normality cannot be assumed for these counts. An appropriate Wilcoxon procedure is therefore to be used to test whether the two word-processors have, on the whole, the same ability in coping with such expressions.

(i) Carry out the test at the 10% level of significance, using the appropriate table in the Students' Handbook.

(ii) Find the actual level of significance of the data as given by the normal approximation to the distribution of the Wilcoxon statistic under the null hypothesis.

You are reminded that the parameters of this normal approximation are

$$\mu = \frac{n(n+1)}{4} \text{ and } \sigma^2 = \frac{n(n+1)(2n+1)}{24}.$$

[MEI]

13 Random samples of size 4 are taken from each of two independent populations. The Wilcoxon rank sum test is to be used to test the hypotheses:

H_0: The two populations have identical distributions.

H_1: The two populations have identical distributions except that their location parameters differ.

The usual procedure of ranking the complete set of eight observations in ascending order is followed (it may be assumed that no two observations are exactly equal). The sum of the ranks corresponding to the observations from one of the populations is denoted by T.

(i) State the minimum and maximum possible values of T.

(ii) State the total number of ways of assigning four ranks from the eight available to the observations from which T is calculated.

(iii) Deduce that the probability that T equals 10 or 26 if H_0 is true is $\frac{2}{70}$.

(iv) Hence *write down* the level of significance of the test that rejects H_0 if the value of T is 10 or 26.

(v) Find the level of significance of the test that rejects H_0 if the value of T is 10, 11, 25 or 26.

(vi) Repeat **(v)** using the normal approximation to the distribution of T under H_0, making appropriate use of a continuity correction. You are reminded that the parameters of this normal approximation are
$\mu = \frac{1}{2}m(m+n+1)$ and
$\sigma^2 = \frac{1}{12}mn(m+n+1)$.

(vii) In a particular experiment, the numerical values of the data are as follows.

Sample 1:	11.4	13.6	10.8	11.9
Sample 2:	12.1	14.4	15.7	14.1

Test H_0 against H_1 at the *six* percent level of significance.

[MEI]

THE WILCOXON SINGLE SAMPLE TEST

1 This is used for testing the null hypothesis that the population median of a random variable is equal to a given value M, under the assumption that the variable is symmetrically distributed about its median.

2 Given a sample, remove any element equal to M. Let n be the size of the reduced sample.

3 To calculate the test statistic:

- Find the absolute differences between each sample value and M.

- Rank these values giving the lowest rank to the smallest difference.

- W_+ and W_- are the sums of ranks of sample values respectively greater or less than M.

- Take the test statistic W to be the smaller of W_- and W_+ for a two-tailed test, or make the appropriate choice for a one-tailed test.

4 The null hypothesis is rejected if W is less than or equal to the appropriate critical value.

5 If n is large W is approximately distributed as

$$N\left(\frac{n(n+1)}{4}, \frac{n(n+1)(2n+1)}{24}\right).$$

THE WILCOXON RANK SUM TEST

1 This is used for testing the null hypothesis that two random variables, X and Y, have the same median under the assumption that their distributions have the same shape.

2 To calculate the test statistic from a sample of m values of X and n values of Y, with $m < n$:

- Rank the sample values from 1 to $(m + n)$, giving the lowest rank to the smallest value.

- W_X and W_Y are the sums of ranks of X values and Y values respectively.

- $T_X = W_X - \frac{1}{2}m(m+1)$ and $T_Y = W_Y - \frac{1}{2}n(n+1)$.

- Take the test statistic T to be the smaller of T_X and T_Y for a two-tailed test, or make the appropriate choice for a one-tailed test.

3 The null hypothesis is then rejected if T is less than or equal to the appropriate critical value.

4 If n is large T is approximately distributed as $N\left(\frac{mn}{2}, \frac{mn(m+n+1)}{12}\right)$.

7 Estimation

Depend upon it, a lucky guess is never merely luck — there's always some talent in it.

Jane Austen

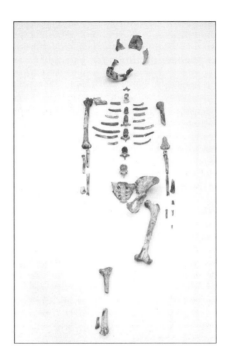

This fossil, known as Lucy, was discovered in Ethiopia between 1973 and 1977 and is widely accepted as the earliest link in the human record. It is 3.3 million years old.

Recently, the remains of a primitive human were found in the Great Rift Valley of Africa. It was the third specimen of its particular subspecies of human. The three specimens found so far are calculated to have been aged 9, 16 and 35 at the times of their deaths.

What can be said about the distribution of ages at death in the population of this species of human, from the sample available?

The obvious answer to this question is 'not much': there is a very small sample to work with, and a very large amount of information is needed to determine the details of a distribution. You need to start by making an heroic assumption about the general shape of the distribution of ages at death. The easiest assumption to start with is that it is a rectangular distribution: humans from this species were equally likely to die at every age up to some maximum. Formally, if X is the random variable 'age at death', then you are assuming that $X \sim \text{Rect}(0, m)$, where m is the maximum age to which these humans lived.

The task is now dramatically simplified. You need to use the sample merely to *estimate the parameter m* of the distribution of ages at death.

❓ What estimate would you make of the parameter m, the maximum age at death, from the sample given?

Estimates for *m*

Two *possible* methods for estimating m are given below, but if you have thought of others, you could follow the work through with your method.

Method 1: Take the largest age in the sample and multiply by $\dfrac{6}{5}$.

Rationale: The largest age in the sample ought to be in the middle of the highest third of the distribution, i.e. at about $\dfrac{5}{6}$ of the maximum age.

Method 2: Find the mean of the sample and double it.

Rationale: The mean age at death ought to be about half of the maximum.

Note that the relationships these methods establish between the experimental outcome and the numerical value produced are referred to as *estimators* for m: the numerical values that the methods produce in particular cases are the *estimates*.

For this sample, these methods give the following estimates.

Method 1: $m_1 = \dfrac{6}{5} \times 35 = 42$

Method 2: $m_2 = 2 \times \left(\dfrac{9 + 16 + 35}{3} \right) = 40$

Estimators as random variables

It is important to realise that an estimator is a random variable – it gives a method for determining a numerical value which cannot be predicted in advance, and which will depend on the sample which happens to be chosen.

In the example above, if you assume that the three sets of remains found were a random sample from the entire population of this subspecies of human, then the ages at death of the three individuals are the values of the three random variables X_1, X_2 and X_3 each of which has the population distribution Rect(0, m).

The variables M_1 and M_2, which are the estimators described by Methods 1 and 2 respectively, are then given by:

$$M_1 = \frac{6}{5}\max\{X_1, X_2, X_3\}$$

and

$$M_2 = \frac{2}{3}(X_1 + X_2 + X_3)$$

As a random variable, an estimator has a distribution. If it is a discrete random variable, the distribution will be the set of possible values which it might take, together with the probability with which each value arises. If it is a continuous random variable, the distribution is given by a density function. The uncertainty involved arises from the process of taking samples at random from the population as a whole, and therefore this distribution is called the *sampling distribution* of the estimator.

In the example, you can derive the sampling distribution of M_1 and M_2 from the distribution of the X_i: this is done below in the case of M_1.

The sampling distribution of M_1

The probability density function for X is:

$$f(x) = \begin{cases} \dfrac{1}{m} & 0 \leqslant x \leqslant m \\ 0 & \text{otherwise} \end{cases}$$

The graph of this function for $m = 40$ is sketched in figure 7.1.

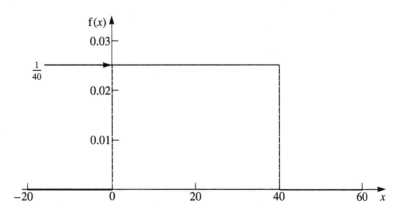

Figure 7.1

The cumulative distribution function for X is obtained by integration.

$$F(x) = P(X \leqslant x) = \int_0^x f(t')\,dt' = \begin{cases} 0 & x \leqslant 0 \\ \dfrac{x}{m} & 0 \leqslant x \leqslant m \\ 1 & x \geqslant m \end{cases}$$

The cumulative distribution function for the estimator M_1 is found by noting that the largest of three numbers will be less than a value x if and only if all three of them are less than x. Thus the probability that the largest of the three sample values is less than x is the same as the probability that all three sample values are less than x.

So you can find the cumulative distribution function of M_1 from the cumulative distribution function of X as follows.

$$F(x) = P(M_1 \leqslant x) = P\left(\max\{X_1, X_2, X_3\} \leqslant \frac{5}{6}x \right) \quad \text{by definition}$$

$$= P\left(X_1 \leqslant \frac{5}{6}x \text{ and } X_2 \leqslant \frac{5}{6}x \text{ and } X_3 \leqslant \frac{5}{6}x \right) \quad \text{by the argument above}$$

$$= P\left(X_1 \leqslant \frac{5}{6}x \right) \times P\left(X_2 \leqslant \frac{5}{6}x \right) \times P\left(X_3 \leqslant \frac{5}{6}x \right) \quad \text{because the } X_i \text{ are independent}$$

$$= \begin{cases} 0 & x \leqslant 0 \\ \left(\dfrac{5x}{6m} \right)^3 & 0 \leqslant x \leqslant \dfrac{6}{5}m \\ 1 & x \geqslant \dfrac{6}{5}m \end{cases}$$

Then the probability density function of M_1, which is found by differentiating the cumulative distribution function, is

$$f(x) = \frac{d}{dx}F(x) = \begin{cases} 3\left(\dfrac{5}{6m} \right)^3 x^2 & 0 \leqslant x \leqslant \dfrac{6}{5}m \\ 0 & \text{otherwise} \end{cases}$$

A graph of this density function is sketched in figure 7.2

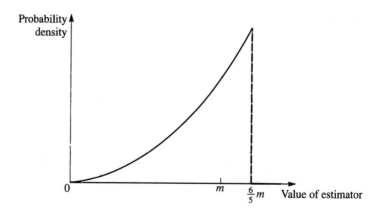

Figure 7.2

This graph illustrates the following points.

- Since each of the X_i is restricted to the range 0 to m, so is the largest of them. Thus the estimator M_1 cannot take values greater than $\frac{6}{5}m$.

- The distribution is strongly negatively skewed: this should not be surprising, since the largest of the X_i is relatively unlikely to be in the lower part of the range 0 to m.

Using a spreadsheet (or a calculator which produces random numbers in the range 0 to 1) you can simulate, for the case $m = 1$, the sampling process used here.

1 Find three random numbers between 0 and 1.
> Simulates measuring X_1, X_2 and X_3

2 Choose the largest.
> Simulates finding max $\{X_1, X_2, X_3\}$

3 Multiply this by $\dfrac{6}{5}$.
> Simulates finding M_1

4 Repeat steps **1**, **2** and **3** 200 times.

5 Plot a histogram showing the simulated sampling distribution of M_1.

It should resemble the probability density function shown in figure 7.2, with $m = 1$.

Good estimators

 What makes a good estimator?

Ideally, whatever the true value of the population parameter, the value taken by the estimator on *every sample* should be *exactly equal* to this value. However, this is unlikely to be possible, because the values in a finite sample could usually arise from more than one possible value of the population parameter, so no calculation with these values can determine the precise value of the parameter.

One way of relaxing this ideal is to ask that, whatever the true value of the population parameter, the value taken by the estimator on *most samples* should be *reasonably near* to this value.

An alternative is to ask that, whatever the true value of the population parameter, the *average value* taken by the estimator over all samples should be *exactly equal* to this value.

Formally, this second criterion means that the expected value of the estimator should be equal to the population parameter: an estimator which has this property is called *unbiased*; other estimators are *biased*.

The motivation behind this criterion comes from considering what happens if the estimator is used repeatedly. In the example, this would mean using it to estimate the lifespans of different subspecies of humans, each time using three specimens of the subspecies. You might expect that a good estimator would sometimes give a value above the true lifespan, and sometimes a value below, but that there should be no average tendency for one of these to predominate. This is what unbiasedness requires. However, you are in fact only going to use this estimator once, so it is not obvious that this criterion is the most sensible one.

In everyday English, to describe something as unbiased is usually to express approval of it: beware of assuming that this is also true in statistics.

For example, consider the estimator, M, for m which is calculated using the method 'pick one of the three specimens at random, and double its age'.

Because the age at death has a rectangular distribution, the specimen picked is equally likely to have any age at death between 0 and m, so that its expected age at death is $\frac{1}{2} m$.

The expected value of M, which is double this, is therefore just m. This means that M is unbiased. However, it should strike you as a rather unsatisfactory way of estimating m from the data.

Are M_1 and M_2 unbiased?

The expected value of M_1 is:

$$E[M_1] = \int_{-\infty}^{\infty} xf(x) \; dx = \int_0^{\frac{6}{5} m} 3 \left(\frac{5}{6m} \right)^3 x^3 \; dx$$

$$= \left[3 \left(\frac{5}{6m} \right)^3 \frac{x^4}{4} \right]_0^{\frac{6}{5} m} = \frac{9m}{10}.$$

This shows that M_1 is biased. It does not, on average over all possible samples, give the true value of the population parameter.

You can see that M_2 is unbiased even though you have not calculated its full sampling distribution. Each of the X_i has the distribution Rect(0, m) and hence $E[X_i] = \frac{1}{2} m.$

Therefore

$$E[M_2] = E\left[\frac{2}{3} (X_1 + X_2 + X_3) \right]$$

$$= E\left[\frac{2}{3} (X_1) \right] + E\left[\frac{2}{3} (X_2) \right] + E\left[\frac{2}{3} (X_3) \right]$$

$$= \frac{2}{3}(E[X_1] + E[X_2] + E[X_3])$$

$$= \frac{2}{3}\left(\frac{m}{2} + \frac{m}{2} + \frac{m}{2}\right) = m.$$

You have found that M_1 is biased, but the estimator:

$$M_3 = \frac{10}{9}M_1 = \frac{4}{3}\max\{X_1, X_2, X_3\}$$

is unbiased, because:

$$E[M_3] = E\left[\frac{10}{9}M_1\right] = \frac{10}{9}E[M_1] = \frac{10}{9} \times \frac{9m}{10} = m.$$

Note

You cannot always create an unbiased estimator from a biased one by multiplying by a constant. It was only possible to do so in this case because the expected value of the biased estimator, although not equal to the population parameter, was just a constant multiple of it.

For the original sample, the value of M_3 is:

$$m_3 = \frac{4}{3} \times 35 = 46\frac{2}{3}.$$

Bias

The *bias* of an estimator for some parameter is simply defined to be the amount by which it is expected to over- or under-estimate that parameter.

That is, if T is an estimator for a parameter τ, the bias of T is

$$\text{bias}(T) = E[T] - \tau$$

so that the bias is positive for an estimator which over-estimates on average, and negative if, on average, it under-estimates. An unbiased estimator is therefore just one with zero bias.

INVESTIGATION

You made the assumption at the start of the analysis that the age at death had a rectangular distribution, but this is not very realistic. In fact, it seems likely that primitive humans were much more likely to die younger. A better model for the distribution of the age at death in such a species might therefore be the triangular distribution with density function:

$$f(x) = \begin{cases} \frac{2}{m^2}(m - x) & 0 \leqslant x \leqslant m \\ 0 & \text{otherwise} \end{cases}$$

The graph of this density function is sketched in figure 7.3.

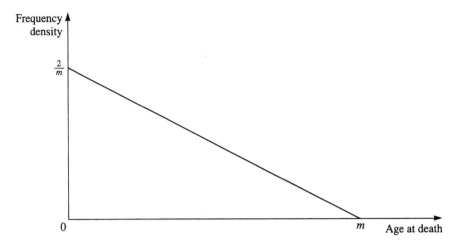

Figure 7.3

Using the same estimators M_1 and M_2 for m and assuming X to have this new distribution, find:

(i) the sampling distribution of M_1

(ii) the expectation of M_1, using your sampling distribution

(iii) the expectation of M_2

and so determine whether M_1 and M_2 are biased or unbiased. If either estimator is biased, find a multiple of it which is unbiased.

Note

In both the original example and in this investigation, the population distribution has been assumed to have a very definite form, so that the determination of the exact population distribution has required the estimation of only one parameter. In general, you may want to make less restrictive assumptions about the form of the population distribution, in which case you may need to estimate two or more parameters. For instance, if you assume that the population distribution is normal, you will want to estimate both its mean and its variance. This example is dealt with in detail later in the chapter.

INVESTIGATION

Use the original (rectangular) distribution for X, but suppose that a sample of size n is available. The estimators M_4 and M_5 are extensions of M_3 and M_2 to this situation.

$$M_4 = \frac{n+1}{n} \, \max\{X_1, X_2, \ldots, X_n\}$$

and:

$$M_5 = \frac{2}{n}(X_1 + X_2 + \ldots + X_n).$$

Determine:

(i) the sampling distribution for M_4

(ii) the expected value of M_4, using your sampling distribution

(iii) the expected value for M_5

and so decide whether M_4 and M_5 are biased or unbiased.

EXAMPLE 7.1

Suppose that you buy scratchcards until you have won three times. The number of cards you have purchased by then is recorded as the value of the random variable N. Calculate the sampling distribution of N and show that $\Pi = \dfrac{2}{N-1}$ is an unbiased estimator of p, the probability that a scratchcard is a winning card.

SOLUTION

The probability that N has the value n is the probability that the third win will occur with the nth card.

That is:

the nth card is a winner

and:

there have been exactly two winners in the first $n - 1$ cards

so

P(third win with nth card) = P(nth card wins) $\times$ P(2 wins in first $n - 1$ cards).

The nth card wins with probability p, and the binomial distribution gives the probability that two of the first $n - 1$ cards are winners (and so $(n - 3)$ are not winners) as:

$$\binom{n-1}{2} p^2 (1 - p)^{(n-3)}.$$

Therefore:

P(third win with nth card) $= p \times \binom{n-1}{2} p^2 (1 - p)^{(n-3)} = \binom{n-1}{2} p^3 (1 - p)^{(n-3)}.$

Note that:

$$\binom{n-1}{2} = \frac{(n-1)(n-2)}{2}.$$

Hence the sampling distribution of N is:

$$P(N = n) = \frac{1}{2}(n - 1)(n - 2)p^3 (1 - p)^{n-3} \quad (n \geqslant 3).$$

(Note that $n < 3$ is impossible – the third win cannot occur before the third card.)

To determine whether $\Pi = \dfrac{2}{N-1}$ is unbiased, you must calculate $E[\Pi]$.

Because Π is a function of N, you can use the result:

$$E[f(N)] = \sum_{\text{all } n} f(n) P(N = n)$$

to obtain:

$$E[\Pi] = \sum_{n=3}^{\infty} \frac{2}{n-1} \times \frac{1}{2}(n-1)(n-2)p^3(1-p)^{n-3}$$

$$= p^3 \sum_{n=3}^{\infty} (n-2)(1-p)^{n-3}.$$

Now substitute $n - 3 = r$. This gives also $n - 2 = r + 1$, and means that $n = 3$ corresponds to $r = 0$ while $n = \infty$ corresponds to $r = \infty$, so $E[\Pi]$ becomes

$$E[\Pi] = p^3 \sum_{r=0}^{\infty} (r+1)(1-p)^r.$$

But this is just a binomial series:

$$(1-a)^{-2} = 1 + (-2)(-a) + \frac{(-2)(-3)}{2!}(-a)^2 + \ldots$$

$$+ \frac{(-2)(-3)\ldots(-[r+1])}{r!}(-a)^r + \ldots$$

$$= 1 + 2a + 3a^2 + \ldots + (r+1)a^r + \ldots$$

$$= \sum_{r=0}^{\infty} (r+1)a^r.$$

So

$$E[\Pi] = p^3(1 - [1-p])^{-2} = p^3 p^{-2}$$

$$= p$$

and Π is unbiased, as required.

EXERCISE 7A

1 A coin is tossed n times and the number of heads, H is counted. The probability of a head is p.

(i) Show that $\Pi = \dfrac{H}{n}$ is an unbiased estimator of p.

(ii) The variance of the underlying binomial distribution is given by $np(1 - p)$. Show that $V = n\Pi(1 - \Pi)$ is not an unbiased estimate of the variance of the distribution. What multiple of V is unbiased?

Hint:
You may find the binomial sum:

$$\sum_{r=0}^{n} r^2 \binom{n}{r} p^r (1-p)^{n-r} = n(n-1)p^2 + np$$

helpful.

2 A typist makes errors at a rate of r per minute. The number of errors, N, he makes in t minutes is counted. Assuming that N is a Poisson variable, write $E[N]$ in terms of r and t.

Show that $R = \dfrac{N}{t}$ is an unbiased estimator of r.

3 An experimenter places six balls, numbered from 1 to 6, in a bag. Three balls are to be drawn from this bag, the largest, L, of the three numbers and the median, M, of the three numbers being noted.

(i) Assume that the balls are drawn without replacement.

(a) Write down all 20 possible samples of size three, and hence calculate the probability distributions for L and M.

(b) Use these distributions to calculate the expected values of L and M.

A subject, who knows that the bag contains n balls numbered 1 to n, but does not know the value of n, is asked to draw three balls from the bag. He uses L and $2M - 1$ as estimates for the value of n.

(c) Explain why $2M - 1$ is unbiased as an estimator for n, but L is not.

(ii) Repeat part (i) for the case where the three balls are drawn with replacement. (There are 56 distinct samples, not all equally likely.)

4 A fairground game requires players to attempt to throw ping-pong balls into pots; the game is so difficult that balls land essentially at random. The pot that wins £1 has twice the area of the pot that wins a teddy-bear, so that, if p is the probability that a ball lands in the teddy-bear pot, $2p$ is the probability that it lands in the £1 pot and $(1 - 3p)$ is the probability that a ball misses the prize pots altogether. To obtain an estimator Π of p, throw two ping-pong balls and see where they land; then record the value of Π, as in this table.

If balls land:	Value of Π
both in one of the prize pots	$\dfrac{1}{3}$
exactly one in one of the prize pots	$\dfrac{1}{6}$
both missing the prize pots	0

(i) Calculate, in terms of p, the probability of each of these outcomes.

(ii) Hence show that Π is an unbiased estimator of p.

(iii) I suspect that the true value of p is about 0.01. How can estimating probabilities as in the table give an unbiased estimate – are they not much too large? Is this a sensible estimation scheme? Can you find a better one based on two balls, or is the crude scheme above the best you can do with so little evidence?

5 The manufacturer of my breakfast cereal is giving away little plastic dinosaurs. One dinosaur is supposed to be included in each packet of cereal but in fact, because of packing difficulties, a certain proportion of packets have – very disappointingly – had their little dinosaurs missed out. I am hoping to estimate the proportion, p, of packets without dinosaurs by waiting until the first packet without a dinosaur turns up. If I have, by then, bought and opened N packets of cereal, I might use the estimator $\Pi = \dfrac{1}{N}$ for p.

(i) Explain why (provided $p \neq 0$ or 1) the distribution of N is geometric:
$$P(N = n) = (1 - p)^{n-1} p.$$

(ii) Check that $E[N] = \dfrac{1}{p}$ so that N is an unbiased estimator for $\dfrac{1}{p}$ ($p \neq 0$ or 1).

(iii) Show, using the logarithm series
$$\sum_{r=1}^{\infty} \frac{x^r}{r} = -\ln(1 - x) \quad (0 \leqslant x < 1),$$
that:
$$E[\Pi] = \frac{p}{1-p} \sum_{n=1}^{\infty} \frac{(1-p)^n}{n} = \frac{p}{1-p} \ln\left(\frac{1}{p}\right)$$
$$(p \neq 0 \text{ or } 1)$$
so that Π is not, in general, unbiased.

(iv) Sketch a graph of $E[\Pi]$ against p $(0 < p < 1)$.

Compare $E[\Pi]$ with p. How bad is the bias? Are there values of p for which this estimate is unbiased?

(v) What happens if $p = 0$ or 1?

6 A coin is tossed repeatedly until h heads have appeared, and the number, N, of tosses required to do this is recorded.

(i) Explain why, if p is the probability of a head: $P(N = n) = P(\text{exactly } h - 1 \text{ heads in the first } n - 1 \text{ tosses and a head on the } n\text{th toss})$ and hence show that:

$$P(N = n) = \binom{n-1}{h-1} p^h (1 - p)^{n-h}$$

$(n = h, h + 1, \ldots)$ is the distribution of N.

(ii) Show that $\Pi = \dfrac{h - 1}{N - 1}$ is an unbiased estimator of p.

(*Hint*: You will need to recognise a binomial sum.)

7 Faced with 100 square kilometres of forest, you are asked to estimate the proportion of diseased trees in the forest. Devise a sampling method and an appropriate estimator, based on the sample. You should bear in mind that some parts of the forest may be less healthy than others.

8 (i) If the n random variables $X_1, X_2, \ldots, X_n$, represent independent drawings from a population whose distribution has mean μ, show that the sample mean:

$$\overline{X} = \frac{1}{n} \sum_{i=1}^{n} X_i$$

is an unbiased estimator of μ.

(ii) More generally, show that if $\alpha_1, \alpha_2, \ldots, \alpha_n$ are any n numbers with the property that:

$$\sum_{i=1}^{n} \alpha_i = 1$$

then the weighted mean

$$\overline{X}_\alpha = \sum_{i=1}^{n} \alpha_i X_i$$

is an unbiased estimator of μ.

(iii) The results above show that, given a random sample of size n drawn from a population of which the distribution has mean μ, then:

$$T_1 = X_1$$

$$T_2 = \frac{1}{5}(4X_1 - 2X_2 + 3X_3)$$

$$T_3 = \frac{1}{n}(X_1 + X_2 + \ldots + X_n)$$

are all unbiased estimators for μ. Which of them is preferable and why? (You might try a computer simulation.)

9 $X_1, X_2, \ldots, X_n$ $(n > 1)$ are independent random variables all having the same normal distribution; σ^2 denotes the common variance. The random variable Y is defined by

$$Y = \sum_{i=1}^{n} (X_i - \overline{X}^2) \text{ where } \overline{X} = \frac{1}{n} \sum_{i=1}^{n} X_i.$$

You may in this question assume the results

$$E[Y] = (n - 1)\sigma^2$$

$$\text{Var}(Y) = 2(n - 1)\sigma^4.$$

(i) It is proposed to estimate σ^2 by an estimator of the form $T = kY$ where k is a constant to be determined. Write down the mean and variance of T.

(ii) The bias of T is defined to be $E[T] - \sigma^2$. Deduce that T will have a zero bias if and only if, $k = \dfrac{1}{(n - 1)}$.

The mean square error of T, mse(T), is given by

$$\text{mse}(T) = \text{Var}(T) + (\text{bias of } T)^2.$$

This is one way of assessing how 'good' T is as an estimator of σ^2, and it is desirable for it to be as small as possible.

(iii) Show that
$$\text{mse}(T) = \sigma^4(n^2 - 1)k^2 - 2\sigma^4(n-1)k + \sigma^4$$

(iv) Hence use calculus to show that the value of k that minimises $\text{mse}(T)$ is $\dfrac{1}{(n+1)}$.

10 A library has been given three books belonging to a very rare set. These books carry volume numbers $X_1, X_2,$ and X_3 (where $X_1 < X_2 < X_3$), but it is not known how many volumes there are altogether in the set.

Suppose that there are n volumes, numbered $1, 2, \ldots, n$, in the set and that the three books in the library are regarded as a random sample from this total of n. Two estimators of n are proposed.

$Y = X_3 +$ average gap between the observed numbers $= \frac{1}{2}(3X_3 - X_1)$

$Z =$ twice sample median $- 1 = 2X_2 - 1$

(i) Consider the case where n is in fact 3. Show that the value of Y is certain to be 4 and that the value of Z is certain to be 3.

(ii) Consider the case where n is in fact 4. Show that the possible values of Y are 4, 5 and $5\frac{1}{2}$.

Show that the mean of Y is 5 and that the variance of Y is $\frac{3}{8}$.

(iii) Still considering the case where n is in fact 4, find the possible values of Z and the mean and variance of Z.

(iv) If n is in fact 5, the following results may be derived:

mean of $Y = 6$ variance of $Y = 0.9$

mean of $Z = 5$ variance of $Z = 2.4$.

(You are *not* required to verify these results.)

Using this information and the results you have obtained in parts (i), (ii) and (iii), discuss, with reasons, whether either Y or Z is preferable as an estimator of n, for the values of n considered in this question.

11 The continuous random variable X has the rectangular distribution on $(0, \theta)$. The following facts may be used without proof in this question:

(A) The p.d.f. of X is $f(x) = \dfrac{1}{\theta}$ for $0 \leqslant x \leqslant \theta$.

(B) The cumulative distribution function of X, i.e. $P(X < x)$, is $f(x) = \dfrac{x}{\theta}$ for $0 \leqslant x < \theta$.

(C) The mean of X is $\dfrac{\theta}{2}$.

(D) The variance of X is $\dfrac{\theta^2}{12}$.

$X_1, X_2, \ldots, X_n$ are independent random variables each distributed as X. The random variable $\overline{X}$ is defined by
$$\overline{X} = \frac{1}{n}\sum_{i=1}^{n} X_i$$
and the random variable Y is defined to be the maximum of $X_1, X_2, \ldots, X_n$.

(i) Use **(C)** and **(D)** above to show that $2\overline{X}$ is an unbiased estimator of θ with variance $\dfrac{\theta^2}{3n}$.

(ii) Deduce from **(B)** that the cumulative distribution function of Y, i.e.
$$P(Y \leqslant y), \text{ is } \left(\frac{y}{\theta}\right)^n \text{ for } 0 \leqslant y \leqslant \theta.$$

(iii) Hence deduce that the p.d.f. of
$$Y \text{ is } \frac{ny^{n-1}}{\theta^n} \text{ for } 0 \leqslant y \leqslant \theta.$$

(iv) Obtain the mean and variance of Y.

(v) Deduce that $Z = \left(1 + \dfrac{1}{n}\right)Y$ is an unbiased estimator of θ and find its variance.

(vi) Would you prefer $2\overline{X}$ or Z as an estimator of θ? Justify your answer.

12 X_1 and X_2 are independent random variables. They both have the same mean μ. Their variances are σ_1^2 and σ_2^2, respectively, where σ_1^2 and σ_2^2 are assumed to be known constants.

It is proposed to estimate μ by an estimator T having the form

$$T = c_1X_1 + c_2X_2$$

where c_1 and c_2 are constants to be determined.

(i) Show that T will be an unbiased estimator of μ if $c_1 + c_2 = 1$. For the remainder of the question, assume this condition holds.

(ii) Find an expression for $\mathrm{Var}(T)$ in terms of c_1, σ_1^2 and σ_2^2.

Hence obtain the value of c_1 that minimises $\mathrm{Var}(T)$. Write down the corresponding value of c_2. Find the minimum of $\mathrm{Var}(T)$.

(iii) Summarise the desirable properties possessed by T (with the obtained values of c_1 and c_2) as an estimator of μ. In practice, what difficulty would arise in estimating μ in this way?

13 $X_1, X_2, \ldots, X_n$ are independent random variables each of which takes value 1 with probability p $(0 < p < 1)$ and value 0 with probability $1 - p$. Let $Y = \sum_{i=1}^{n} X_i$ so that

$$\overline{X} = \frac{1}{n}\sum_{i=1}^{n} X_i = \frac{1}{n}Y.$$ Using the fact that Y is a

$B(n, p)$ random variable,

(i) show that $\overline{X}$ is an unbiased estimator of p

(ii) show that $\mathrm{Var}(\overline{X}) = \dfrac{p(1-p)}{n}$

(iii) show that $\mathrm{E}\left[\dfrac{\overline{X}(1-\overline{X})}{n}\right] = \dfrac{n-1}{n}\dfrac{p(1-p)}{n}$

(iv) deduce the constant k such that $k\overline{X}(1 - \overline{X})$ is an unbiased estimator of $\mathrm{Var}(\overline{X})$.

[MEI]

14 The number of breakdowns per week on a computer network is a random variable Y

having the Poisson distribution with parameter θ. State the mean and variance of Y, and hence find $\mathrm{E}[Y^2]$.

The random variables $Y_1, Y_2, \ldots, Y_n$ represent a random sample of observations of the weekly number of breakdowns. Show that $\overline{Y}$ is an unbiased estimator of θ and state its variance.

Find the function of θ for which $\dfrac{1}{n}\sum Y_i^2$ is an unbiased estimator.

The weekly cost of repairing the breakdowns is modelled by

$$C = 4Y + 2Y^2.$$

Show that $\mathrm{E}[C] = 6\theta + 2\theta^2$. Find constants a and b such that $a\overline{Y} + \dfrac{b}{n}\sum Y_i^2$ is an unbiased estimator of $\mathrm{E}[C]$.

[MEI]

15 A sequence of independent trials, each of which has two outcomes (success or failure), is carried out. On each trial, the probability of a success is p $(0 < p < 1)$ and that of a failure is $q = 1 - p$. The random variable X counts the number of trials up to and including the first success.

(i) Explain why $P(X = x) = q^{x-1}p$ for $x = 1, 2, \ldots$.

(ii) Show that $\mathrm{E}[X] = \dfrac{1}{p}$.

(iii) Deduce that a reasonable estimator of p is $\dfrac{1}{\overline{X}}$.

(iv) Find $\mathrm{E}\left[\dfrac{1}{X}\right]$. Determine whether or not $\dfrac{1}{\overline{X}}$ is an unbiased estimator of p.

(v) Given that $\mathrm{Var}(X) = \dfrac{1-p}{p^2}$, find $\mathrm{E}[X^2]$ and hence find constants a and b such that $aX^2 + bX$ is an unbiased estimator of $\mathrm{Var}(X)$.

(You may find it helpful to use the results

$$(1 - q)^{-2} = 1 + 2q + 3q^2 + \ldots$$

$$\ln(1 - q) = -q - \frac{1}{2}q^2 - \frac{1}{3}q^3 - \ldots$$

for $0 < q < 1$.)

Estimators for population mean and variance

In *Statistics 3*, you saw how to estimate the mean and variance of a population from a sample. Now that you have looked at the idea of an estimator, you can derive these results more formally.

Consider the following model. You take a random sample of size n from a population which is very large (so that you can consider sampling with or without replacement to be identical). You measure, for each element of the sample, a variable, X, for which the distribution in the population has mean μ and variance σ^2, which are unknown parameters. Taking the sample can be considered as measuring the values of the n random variables $X_1, X_2, \ldots, X_n$ each of which has mean μ and variance σ^2 and which are all independent of one another, because of the assumption that the population is very large. The problem is to construct from $X_1, X_2, \ldots, X_n$ random variables which are estimators for μ and σ^2.

An unbiased estimator for μ

An obvious candidate for an estimator for μ is:

$$M = \frac{X_1 + X_2 + \ldots + X_n}{n}$$

the sample mean. To show that M is unbiased, calculate its expectation.

$$E[M] = E\left[\frac{X_1 + X_2 + \ldots + X_n}{n}\right] = E\left[\frac{1}{n}\{X_1 + X_2 + \ldots + X_n\}\right]$$

$$= \frac{1}{n}E[X_1 + X_2 + \ldots + X_n] \quad \text{(because } E[tA] = tE[A] \text{ for constant } t\text{)}$$

$$= \frac{1}{n}\{E[X_1] + E[X_2] + \ldots + E[X_n]\}$$

$$\text{(because } E[A + B + \ldots] = E[A] + E[B] + \ldots\text{)}$$

$$= \frac{1}{n}\{\mu + \mu + \ldots + \mu\} = \mu$$

So M is an unbiased estimator of μ.

An estimator for σ^2

You might naturally try, as an estimator of σ^2, the variance of the n elements of the sample. Define V as:

$$V = \frac{(X_1 - M)^2 + (X_2 - M)^2 + \ldots + (X_n - M)^2}{n}$$

then, multiplying out all the brackets and collecting the similar terms:

$$V = \frac{1}{n}\{(X_1^2 - 2X_1M + M^2) + (X_2^2 - 2X_2M + M^2) + \ldots + (X_n^2 - 2X_nM + M^2)\}$$

$$= \frac{X_1^2 + X_2^2 + \ldots + X_n^2}{n} - \frac{2M(X_1 + X_2 + \ldots + X_n)}{n} + \frac{M^2 + M^2 + \ldots + M^2}{n}$$

$$= \frac{X_1^2 + X_2^2 + \ldots + X_n^2}{n} - 2M\left(\frac{X_1 + X_2 + \ldots + X_n}{n}\right) + M^2$$

$$= \frac{X_1^2 + X_2^2 + \ldots + X_n^2}{n} - 2M^2 + M^2 \qquad \left(\text{because } \tfrac{X_1 + X_2 + \ldots + X_n}{n} = M\right)$$

$$= \frac{X_1^2 + X_2^2 + \ldots + X_n^2}{n} - M^2.$$

So, again using the results $E[tA] = tE[A]$ and $E[A + B + \ldots] = E[A] + E[B] + \ldots$:

$$E[V] = \frac{1}{n}\{E[X_1^2] + E[X_2^2] + \ldots + E[X_n^2]\} - E[M^2]$$

$$= E[X^2] - E[M^2]. \qquad \left(\text{because } E[X_1^2] = E[X_2^2] = \ldots = E[X^2]\right)$$

But you know that:

$$\text{Var}[R] = E[R^2] - (E[R])^2$$

for any random variable R, and so, turning this around:

$$E[R^2] = \text{Var}[R] + (E[R])^2.$$

Hence $E[V] = \text{Var}[X] + \mu^2 - [\text{Var}[M] + \mu^2] = \text{Var}[X] - \text{Var}[M]$.

You can also obtain the familiar result

$$\text{Var}[M] = \text{Var}\left[\frac{1}{n}\{X_1 + X_2 + \ldots + X_n\}\right]$$

$$\qquad \left(\text{because } \text{Var}[tA] = t^2\text{Var}[A] \text{ for constant } t\right)$$

$$= \frac{1}{n^2}\text{Var}[X_1 + X_2 + \ldots + X_n]$$

$$= \frac{1}{n^2}\{\text{Var}[X_1] + \text{Var}[X_2] + \ldots + \text{Var}[X_n]\}$$

$$\qquad \left(\begin{array}{l}\text{because } \text{Var}[A + B + \ldots] = \text{Var}[A] + \text{Var}[B] + \ldots \\ \text{for independent } A, B, \ldots\end{array}\right)$$

$$= \frac{1}{n^2}\{\sigma^2 + \sigma^2 + \ldots + \sigma^2\} = \frac{\sigma^2}{n}.$$

Thus:

$$E[V] = \text{Var}[X] - \text{Var}[M]$$

$$= \sigma^2 - \frac{\sigma^2}{n}$$

$$= \frac{n-1}{n}\sigma^2.$$

Therefore V is not an unbiased estimator of σ^2. On average, in fact, V will under-estimate σ^2. However, the estimator S^2 which is obtained by multiplying V by $\dfrac{n}{n-1}$ is unbiased:

$$S^2 = \frac{n}{n-1}V = \frac{\sum_{i=1}^{n}(X_i - M)^2}{n-1}.$$

It gives the sample estimate of the population variance which you used in *Statistics 3* with the *t*-distribution.

$$s^2 = \frac{\sum_{i=1}^{n}(x_i - \bar{x})^2}{n-1}$$

Remember that this can also be written in the form:

$$s^2 = \frac{\sum_{i=1}^{n}x_i^2 - n \cdot \bar{x}^2}{n-1}$$

which is sometimes more convenient.

Note

Recall that this estimate of the variance is said to have $(n-1)$ *degrees of freedom*. Intuitively this is because one of the n independent sources of variability in the sample is 'used up' in estimating the mean – you are calculating the variance of the sample data about the sample mean, which is the value which minimises this variance, rather than about the true mean.

Pooled samples

My local grocer has a set of scales to which he gives a quick check every morning by weighing one packet from that morning's delivery of butter from the dairy, which he reckons varies in weight only a little. Over eight days, the weights (in grams) recorded by his scales are:

$$407, \ 417, \ 416, \ 402, \ 409, \ 411, \ 409, \ 413$$

with mean, 410.5 g.

From these eight readings you can estimate:

$$s_1^2 = \frac{\sum\limits_{i=1}^{8} x_i^2 - 8\bar{x}^2}{8 - 1} = \frac{1\,348\,250 - 8\times(410.5)^2}{7} = 24.$$

Last week, the service engineer from the company which leases him his scales came in to reset them, which he does every three months. Over the next five days, the weights (in grams) of his checking packets of butter are recorded as:

$$423, \ 415, \ 418, \ 423, \ 427$$

which seem to be higher. They have mean 421.2 g.

From these five readings you can estimate:

$$s_2^2 = \frac{\sum\limits_{i=1}^{5} y_i^2 - 5\bar{y}^2}{5 - 1} = \frac{887\,136 - 5\times(421.2)^2}{4} = 22.2.$$

You can take it that there has been a shift in the measured weights of the butter because of the adjustment of the scales, but that the variance of the measurements, presumably caused by the small daily variation in the butter weights, will have stayed the same. How could you estimate this common variance?

The statistical model of the grocer's sampling process is that he has taken a sample of size $n_1 = 8$ from a population of butter packets with unknown mean μ_1 and unknown variance σ^2; then a sample of size $n_2 = 5$ from a population of butter packets with unknown and (possibly) different mean μ_2 and unknown but identical variance σ^2.

In this case, an appropriate unbiased estimate of σ^2 is given by:

$$s^2 = \frac{\sum\limits_{i=1}^{n_1}(x_i - \bar{x})^2 + \sum\limits_{i=1}^{n_2}(y_i - \bar{y})^2}{(n_1 + n_2 - 2)} = \frac{(n_1 - 1)s_1^2 + (n_2 - 1)s_2^2}{(n_1 + n_2 - 2)}.$$

Before you prove this, make sure you can see why it is intuitively sensible. Figure 7.4 shows what the underlying distributions of butter weights before and after adjustment might look like: you are trying to find the common spread of these distributions, so that it makes sense to look at how far each weight is from the centre of the appropriate distribution (which you estimate by the separate sample means), not how far it is from some common mean value somewhere in between the two distributions.

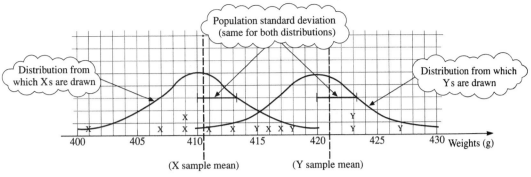

Population standard deviation
(same for both distributions)

Distribution from
which Xs are drawn

Distribution from which
Ys are drawn

400   405   410   415   420   425   430 Weights (g)

(X sample mean)   (Y sample mean)

(Xs and Ys indicate sample values)

Figure 7.4

This explains the numerator; the denominator fits what was said about degrees of freedom above – two of the $n_1 + n_2$ independent pieces of information in the sample are contained in the two sample means and the remaining $n_1 + n_2 - 2$ independent pieces of information are used to determine the variance.

PROOF

Let S^2, S_1^2, S_2^2 denote the estimators of which s^2, s_1^2, s_2^2, are particular values. Then:

$$E[S^2] = E\left[\frac{(n_1 - 1)S_1^2 + (n_2 - 1)S_2^2}{(n_1 + n_2 - 2)}\right]$$

$$= E\left[\frac{(n_1 - 1)}{(n_1 + n_2 - 2)}S_1^2 + \frac{(n_2 - 1)}{(n_1 + n_2 - 2)}S_2^2\right].$$

But $E[aX + bY] = aE[X] + bE[Y]$ for constant a and b so

$$E[S^2] = \frac{(n_1 - 1)}{(n_1 + n_2 - 2)}E[S_1^2] + \frac{(n_2 - 1)}{(n_1 + n_2 - 2)}E[S_2^2]$$

$$= \frac{(n_1 - 1)}{(n_1 + n_2 - 2)}\sigma^2 + \frac{(n_2 - 1)}{(n_1 + n_2 - 2)}\sigma^2$$

because both S_1^2 and S_2^2 are unbiased estimates of σ^2, so:

$$E[S^2] = \frac{(n_1 - 1) + (n_2 - 1)}{n_1 + n_2 - 2}\sigma^2 = \sigma^2.$$

An unbiased estimate of the variance of the weights of butter packs is therefore:

$$s^2 = \frac{7 \times 24 + 4 \times 22.2}{8 + 5 - 2} = \frac{256.8}{11} = 23.3.$$

In general, the process of calculating a pooled estimate of the variance is used where a statistical model identifies two samples as coming from populations with different means but the same variance: the pooled-sample estimate is then an unbiased estimate of that common variance.

EXERCISE 7B

1 Use the data below to calculate unbiased estimates of the population variance.

(i) A single random sample

(a) 12.3, 17.1, 5.7, 12.1, 10.3, 14.2, 6.9, 7.1, 10.7

(b) $n = 23$, $\sum_{i=1}^{23} x_i = 1782.4$,

$$\sum_{i=1}^{23} x_i^2 = 141\,328.6$$

(ii) Two random samples drawn from populations with the same variance, but possibly different means

(a) First sample: 23.2, 31.6, 28.4, 27.9, 29.0, 33.1, 27.6, 30.2

Second sample: 33.2, 31.4, 35.7, 36.8, 33.2

(b) First sample:

$$n = 17, \sum_{i=1}^{17} x_i = 451.88,$$

$$\sum_{i=1}^{17} x_i^2 = 12\,186.34$$

Second sample:

$$n = 14, \sum_{i=1}^{14} y_i = 439.72,$$

$$\sum_{i=1}^{14} y_i^2 = 14\,204.67$$

(c) First sample:

$$n = 11, \bar{x} = 250.4, s_x^2 = 17.340$$

Second sample:

$$n = 7, \bar{y} = 213.2, s_y^2 = 26.031$$

2 An experiment requires a balloon to be dropped from three different heights and the time of descent from each height to be measured. A student conducting this experiment decides to time the descent from each height several times to get a more accurate result. The error made in each time measurement can be modelled as normally distributed with a mean which may have a different non-zero value for each height but a constant variance. His times, in seconds, are listed below.

First height: 7.23, 7.18, 7.19, 7.25, 7.13
Second height: 5.61, 5.71, 5.70, 5.58, 5.61, 5.63
Third height: 3.82, 3.66, 3.83, 3.86

Calculate an unbiased estimate of the variance of the error from these data (you will have to extend the ideas in the text to three samples).

3 Two samples are taken from distributions with the same variance but with means which may differ. Unbiased estimates of the variance are derived as follows:

From the first sample: 28.375

From the second sample: 30.86

From the combined sample: 30.15

Given that the combined sample had 16 elements, calculate the sizes of the two smaller samples.

4 Suppose two separate samples $x_1, x_2, \ldots, x_{n_1}$ and $y_1, y_2, \ldots, y_{n_2}$ have been taken from the *same population*, and you want to pool the samples for the purposes of estimating the population variance. There is no problem, of course, if you have all the data available to you, but suppose the information you have is limited to the estimates of the population mean and variance derived separately from the two samples. That is, you know $\bar{x}, s_x^2, n_1$ and $\bar{y}, s_y^2, n_2$ and you need to determine s^2.

7 Estimation

(i) Calculate $\sum x$ and $\sum x^2$ in terms of $\bar{x}, s_x^2$ and n_1 and $\sum y$ and $\sum y^2$ in terms of $\bar{y}, s_y^2$ and n_2. Let $z_1, z_2, \ldots, z_{n_1+n_2}$ describe the pooled sample so that:

$$\sum z = \sum x + \sum y;$$

$$\sum z^2 = \sum x^2 + \sum y^2.$$

(ii) Hence calculate s^2.

(iii) Explain how this process differs from that by which a pooled variance estimate was derived in the text.

5 The random variables $X_1, X_2, \ldots, X_n$ (where $n > 1$) are independent and they all have the same distribution with mean μ and variance σ^2. The random variable $\bar{X}$ is defined by

$$\bar{X} = \frac{1}{n}\sum_{i=1}^{n} X_i.$$

(i) Show that

$$\sum_{i=1}^{n}(X_i - \bar{X})^2 = \sum_{i=1}^{n}(X_i - \mu)^2 - n(\bar{X} - \mu)^2.$$

(ii) Write down the expected value of

$$\sum_{i=1}^{n}(X_i - \mu)^2.$$

You are reminded that, if the random variable U has mean θ,

$$\text{Var}(U) = E[(U - \theta)^2].$$

(iii) State the variance of $\bar{X}$ and hence write down the expected value of $n(\bar{X} - \mu)^2$.

(iv) Deduce that

$$Y = \frac{1}{n-1}\sum(X_i - \bar{X})^2$$

is an unbiased estimator of σ^2.

(v) You are given the result that if the common distribution of the X_i is $N(\mu, \sigma^2)$, then Y has variance $\dfrac{2\sigma^4}{n-1}$ and, if n is large, is approximately normally distributed.

Using this result explain why, in large samples, an approximate 95% confidence interval for σ^2 is given by

$$s^2\left(1 \pm 1.96\sqrt{\frac{2}{n-1}}\right)$$

where s^2 is the observed value of

$$\frac{1}{n-1}\sum(X_1 - \bar{X}^2).$$

6 You may in this question *use* the result that

$$\int_0^\infty y^m e^{-y}\, dy = m!$$

for non-negative integer m.

(i) The probability density function of the continuous random variable X is

$$f(x) = \begin{cases} k(x-\theta)^{v-1}e^{-(x-\theta)} & \theta < x < \infty \\ 0 & \text{elsewhere} \end{cases}$$

where the parameters θ and v are both positive and v is an integer. Show, by using the substitution $y = x - \theta$, that $k = \dfrac{1}{(v-1)!}$

(ii) Show that the mean of X is $\theta + v$.

(iii) Find $E[X^2]$ and hence show that the variance of X is v.

(iv) Deduce that plausible estimators of θ and v are $\bar{X} - S^2$ and S^2. where $\bar{X}$ and S^2 are defined as usual by

$$\bar{X} = \frac{1}{n}\sum_{i=1}^{n} X_i$$

$$S^2 = \frac{1}{n-1}\sum_{i=1}^{n}(X_i - \bar{X})^2$$

and $\{X_1, X_2, \ldots, X_n\}$ is a random sample of observations on X.

(v) State, with *brief* justifications, whether or not S^2 is an unbiased estimator of v and whether or not $\bar{X} - S^2$ is an unbiased estimator of θ.

7 (i) The random variable Y has mean α. The variance of Y is defined by $\mathrm{Var}(Y) = \mathrm{E}[(Y - \alpha)^2]$; prove from this definition that $\mathrm{Var}(Y) = \mathrm{E}[Y^2] - \alpha^2$.

(ii) Deduce that $\mathrm{E}[Y^2] = \mathrm{Var}(Y) + (\mathrm{E}[Y])^2$.

(iii) The random variable X has a distribution with mean μ and variance σ^2, *where μ is known*. Use the result in **(ii)** to *write down* an expression for $\mathrm{E}[X^2]$ in terms of μ and σ^2.

(iv) $X_1, X_2, \ldots, X_n$ are independent random variables each with the same distribution X in **(iii)**. Using the definition of variance in **(i)**, or otherwise, show that

$$\frac{1}{n}\sum_{i=1}^{n}(X_i - \mu)^2$$

is an unbiased estimator of σ^2.

(v) The random variable $\overline{X}$ is defined by $\overline{X} = \frac{1}{n}\sum_{i=1}^{n}X_i$. Use the result in **(ii)** to *write down* an expression for $\mathrm{E}[\overline{X}^2]$ in terms of μ, σ^2 and n.

(vi) The random variable T is defined by

$$T = \frac{1}{n}\sum_{i=1}^{n}(X_i - \overline{X})^2.$$

Show that

$$T = \frac{1}{n}\sum_{i=1}^{n}X_i^2 - \overline{X}^2.$$

(vii) Deduce that T is not an unbiased estimator of σ^2.

[MEI]

Methods for comparing estimators

Earlier, you considered what can be said about the distribution of ages at death in the population of this subspecies of human, from the sample available. It was assumed there that the distribution of ages at death is a rectangular distribution: humans from this subspecies were equally likely to die at every age up to some maximum. Formally, if the random variable X is age at death, then it was assumed that $X \sim \mathrm{Rect}(0, m)$, where m is the maximum age to which these humans lived. The sample was then used to estimate the parameter m of the distribution of ages at death.

This section will extend this earlier work, and consider an alternative method of comparing estimators.

The estimator you will consider is

$$G = \alpha \cdot \max\{X_1, X_2, X_3\}$$

for different values of α where $X_1, X_2,$ and X_3 are the random variables representing the three elements of the sample.

The sampling distribution of *G*

The probability density function for X is the rectangular distribution

$$F(x) = \begin{cases} \dfrac{1}{m} & 0 \leqslant x \leqslant m \\ 0 & \text{otherwise.} \end{cases}$$

The cumulative distribution function for X is obtained by integration

$$F(x) = P(X \leqslant x) = \int_{-\infty}^{x} f(t)\, dt = \begin{cases} 0 & x \leqslant 0 \\ \dfrac{x}{m} & 0 \leqslant x \leqslant m \\ 1 & x \geqslant m \end{cases}$$

The cumulative distribution function for the estimator G is found by noting that the largest of three numbers will be less than a value x if, and only if, all three of them are less than x.

So, the probability that the largest of the three sample values is less than x is the same as the probability that all three sample values are less than x.

You can find the cumulative distribution function of G from the cumulative distribution function of X as follows:

$$F(x) = P(G \leqslant x) = P\left(\max\{X_1, X_2, X_3\} \leqslant \frac{x}{\alpha} \right)$$

$$= P\left(X_1 \leqslant \frac{x}{\alpha} \text{ and } X_2 \leqslant \frac{x}{\alpha} \text{ and } X_3 \leqslant \frac{x}{\alpha} \right)$$

$$= P\left(X_1 \leqslant \frac{x}{\alpha} \right) \times P\left(X_2 \leqslant \frac{x}{\alpha} \right) \times P\left(X_3 \leqslant \frac{x}{\alpha} \right)$$

$$= \begin{cases} 0 & x \leqslant 0 \\ \left(\dfrac{x}{\alpha m} \right)^3 & 0 \leqslant x \leqslant \alpha m \\ 1 & x \geqslant \alpha m. \end{cases}$$

Therefore, the probability density function of G, which is found by differentiating the cumulative distribution function, is

$$f(x) = \frac{d}{dx} F(x) = \begin{cases} \dfrac{3x^2}{(\alpha m)^3} & 0 \leqslant x \leqslant \alpha m \\ 0 & \text{otherwise.} \end{cases}$$

A graph of this density function is plotted in figure 7.5 for $m = 50$ and three different values of α.

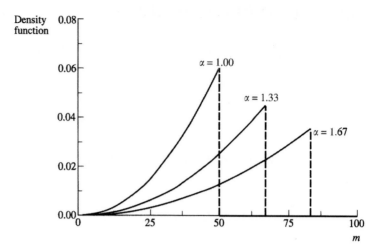

Figure 7.5

You can also use the distribution of G to calculate its mean and variance.

$$E[G] = \int_0^{\alpha m} xf(x)\ dx = \int_0^{\alpha m} x\frac{3x^2}{(\alpha m)^3}\ dx = \left[\frac{3x^4}{4(\alpha m)^3}\right]_0^{\alpha m} = \frac{3\alpha m}{4}$$

$$\text{and } E[G^2] = \int_0^{\alpha m} x^2 f(x)\ dx = \int_0^{\alpha m} x^2\frac{3x^2}{(\alpha m)^3}\ dx = \left[\frac{3x^5}{5(\alpha m)^3}\right]_0^{\alpha m} = \frac{3(\alpha m)^2}{5}$$

$$\text{so } \mathrm{Var}[G] = E[G]^2 - (E[G])^2 = \frac{3(\alpha m)^2}{5} - \left(\frac{3\alpha m}{4}\right)^2 = \frac{3(\alpha m)^2}{80}.$$

The estimator G will be unbiased if α has the value $\frac{4}{3}$, since in this case $E[G] = m$, which is the definition of unbiasedness: the expected value of an estimator is equal to the parameter being estimated. Remember that the intuitive meaning of unbiasedness is that if the estimator G with $\alpha = \frac{4}{3}$, is used many times on samples of size 3 to find the parameter m in situations modelled by the rectangular distribution, it will give the correct result on average.

The bias of G in the cases of $\alpha = 1$ and $\alpha = \frac{5}{3}$ is of the same magnitude (though with opposite sign).

$$\alpha = 1: \quad \mathrm{bias}[G] = E[G] - m = \frac{3m}{4} - m = -\frac{m}{4}$$

$$\alpha = \frac{5}{3}: \quad \mathrm{bias}[G] = E[G] - m = \frac{5m}{4} - m = \frac{m}{4}$$

However, looking at the density functions in figure 7.5 which correspond to these two cases, the estimator with $\alpha = 1$ seems intuitively more satisfactory because, although estimates will be consistently too small, they will only rarely be very far from the true value of 50.

The estimator with $\alpha = \frac{5}{3}$, although only as far from the true value on average, is spread much more widely about this mean and can therefore be expected to take values far from the true value much more frequently. The formal way that you have seen for describing this is that the estimator G has a much larger standard deviation in the $\alpha = \frac{5}{3}$ case than in the $\alpha = 1$ case.

In general, when investigating how good an estimator is, you will want to consider its bias, that is how far its mean is from the true parameter value, but also its standard deviation as a measure of its spread about this mean value. The term *standard error* is usually used to refer to the standard deviation of an estimator.

EXAMPLE 7.2

The variable X has a normal distribution with unknown mean μ and standard deviation σ.

(i) A sample $X_1, X_2, \ldots, X_n$ of size n is taken and the sample mean $\overline{X}$ is used to estimate μ.

(ii) A sample of size 3 is taken, and the estimator $L = \frac{1}{3}(7X_1 - 8X_2 + 4X_3)$ is used to estimate μ.

In each case, show that the estimator is unbiased and find its standard error.

SOLUTION

(i) You know that $\mathrm{E}[\overline{X}] = \mu$ and that $\mathrm{Var}[\overline{X}] = \dfrac{\sigma^2}{n}$ so that the standard error of $\overline{X}_n$ is $\dfrac{\sigma}{\sqrt{n}}$.

Note that the standard error decreases with sample size.

(ii) $\mathrm{E}[L] = \dfrac{1}{3}(7\mathrm{E}[X_1] - 8\mathrm{E}[X_2] + 4\mathrm{E}[X_3])$

$= \dfrac{1}{3}(7\mu - 8\mu + 4\mu) = \mu$

$\mathrm{Var}[L] = \dfrac{1}{3^2}\left(7^2\mathrm{Var}[X_1] + 8^2\mathrm{Var}[X_2] + 4^2\mathrm{Var}[X_3]\right)$

$= \dfrac{1}{9}(49\sigma^2 + 64\sigma^2 + 16\sigma^2) = \dfrac{43\sigma^2}{3}$

so that the standard error of L is $\sqrt{43}\dfrac{\sigma}{\sqrt{3}}$: that is, about 6.5 times as large as that of $\overline{X}$, in the case $n = 3$.

In practice, you would not normally know the value of σ and so you could only estimate the standard error of the sample mean, by making a sample estimate of the variance.

EXAMPLE 7.3

Given that 11, 12, 17, 22, 35 and 47 are six values sampled from a normal distribution with unknown mean μ and unknown standard deviation σ, find an estimate of μ and of the standard error of this estimate.

SOLUTION

From these data:
$$\bar{x} = \frac{1}{6}(11 + 12 + 17 + 22 + 35 + 47) = 24$$

$$s^2 = \frac{6}{5} \times \left(\frac{1}{6}(11^2 + 12^2 + 17^2 + 22^2 + 35^2 + 47^2) - 24^2 \right) = 203.2$$

so an (unbiased) estimate of μ is 24, and the standard error (se) of this sample mean is estimated as $\text{se}(\bar{x}) = \sqrt{\frac{203.2}{6}} = 5.82$.

Mean square error

In trying to capture what you want from a good estimator, you might sum it up as 'not being too far away from the true value too often'. The *mean square error* is the expected squared difference between the estimator and the true value of the parameter being estimated. Asking for an estimator to have a small mean square error then translates this idea of a good estimator fairly directly into formal language. If T is an estimator for parameter τ, then the mean square error is

$$\text{mse}(T) = E[(T - \tau)^2].$$

For the estimator G in the example on page 165 you can calculate the mean square error directly.

$$\text{mse}(G) = E[(G - m)^2] = \int_0^{\alpha m} (x - m)^2 f(x) \, dx = \int_0^{\alpha m} (x - m)^2 \frac{3x^2}{(\alpha m)^3} \, dx$$

$$= \frac{3}{(\alpha m)^3} \left[\frac{x^5}{5} - 2m \frac{x^4}{4} + m^2 \frac{x^3}{3} \right]_0^{\alpha m} = \frac{m^2}{10}(6\alpha^2 - 15\alpha + 10)$$

The minimum value of this expression occurs when $\alpha = \frac{5}{4}$ (by differentiation or completing the square). Note that this is not the value of α which makes G unbiased.

You should be clear that the mean square error is not a unique measure of how good an estimator is. An alternative measure would be the *mean absolute error (mae)*, for instance: mae $(T) = E[|T - \tau|]$.

To choose the mean square error as a criterion is to make a choice about the relative weighting of prospective large and small errors in estimation: the mean square error weights large errors relatively more highly than the mean absolute error would. In the end, the problem comes down to one you have met before: what makes a good estimator depends on what it is to be used for, and how expensive it will be to make mistakes of various sizes or types.

An alternative formula for the mean square error

In the calculation above, the mse for G was relatively easy to find, but in more complex cases an alternative formula is useful.

$$\text{mse}(T) = \text{E}[(T - \tau)^2] \qquad \text{by definition}$$

$$= \text{E}[T^2] - \text{E}[2\tau T] + \text{E}[\tau^2] \qquad \text{expanding and using } \text{E}[X \pm Y] = \text{E}[X] \pm \text{E}[Y]$$

$$= \text{E}[T^2] - 2\tau\text{E}[T] + \tau^2 \qquad \text{since } \tau \text{ is non-random}$$

$$= \text{E}[T^2] - \text{E}[T]^2 + \text{E}[T]^2 - 2\tau\text{E}[T] + \tau^2$$

$$= \text{Var}[T] + (\text{E}[T] - \tau)^2 \qquad \text{using } \text{Var}[T] = \text{E}[T^2] - \text{E}[T]^2$$

Noting that $(\text{E}[T] - \tau)$ measures the bias of T, the extent to which $\text{E}[T]$ is not equal to the true value of the parameter τ, you can summarise this as

$$\text{mse}(T) = \text{Var}[T] + (\text{bias}(T))^2 = (\text{standard error}(T))^2 + (\text{bias}(T))^2.$$

This shows that to have a small mean square error requires a small spread in the values given by the estimator (standard error is small) about a mean which is not too far from the true value (bias is small): this corresponds neatly to the intuitive idea of what constitutes a good estimator.

As an example, look at how the mean square error in G is made up, for the values $\alpha = 1, \frac{5}{4}, \frac{4}{3}$ and with $m = 50$.

α	E[G]	Bias[G] = E[G] – 50	Var[G]	mse[G] = Var[G] + (Bias [G])^2	Comment
1	37.5	−12.5	93.75	250.00	a small spread, but about a mean value which is far from the true parameter
$\frac{5}{4}$	46.875	−3.125	146.48	156.25	a compromise between the two effects: a moderate spread about a mean not too far from the true parameter value
$\frac{4}{3}$	50.0	0.0	166.67	166.67	a mean equal to the true parameter value, but a large spread about the mean

Efficiency

If one estimator of a particular parameter has a smaller mean square error than another for all possible values of the parameter, then the first estimator is said to be a more *efficient* estimator of this parameter. If the two estimators are both unbiased, then the formula above shows that one estimator is more efficient than the other, if it has the smaller standard error for all values of the parameter.

You saw above that the value of α which makes G as efficient as possible is $\alpha = \frac{5}{4}$.

EXAMPLE 7.4

The variable X has a distribution with unknown mean μ and standard deviation s.

A sample X_1, X_2, X_3 of size 3 is taken and the estimator $L = aX_1 + bX_2 + cX_3$ used to estimate μ.

(i) If L is required to be unbiased, find the values of a, b and c which make L as efficient as possible, and state the mean square error in this case.

(ii) Find the mean square error in the case $a = b = c = \frac{1}{4}$. Compare this estimator with the one found in **(i)**.

SOLUTION

(i) $E[L] = E[aX_1 + bX_2 + cX_3] = aE[X_1] + bE[X_2] + cE[X_3] = (a + b + c)\mu$.

For L to be unbiased this must equal μ, so that $(a + b + c) = 1$.

Since L is unbiased,

$$\text{mse}(L) = \text{Var}[L]$$

$$= \text{Var}[aX_1 + bX_2 + cX_3]$$

$$= a^2\text{Var}[X_1] + b^2\text{Var}[X_2] + c^2\text{Var}[X_3]$$

$$= (a^2 + b^2 + c^2)\sigma^2$$

so that the question is asking for the minimum value of $(a^2 + b^2 + c^2)$, given that $(a + b + c) = 1$. Because

$$(a^2 + b^2 + c^2) = \frac{1}{3}\{(a + b + c)^2 + (a - b)^2 + (b - c)^2 + (c - a)^2\},$$

this minimum value occurs when the last three terms of the expression in curly brackets are all equal to zero, i.e. when $a = b = c = \frac{1}{3}$ (so that $(a + b + c) = 1$). In this case, the mean square error is

$$\frac{1}{3}(a + b + c)^2\sigma^2 = \frac{1}{3}\sigma^2.$$

(ii) In the case $a = b = c = \dfrac{1}{4}$,

$$\text{Var}[L] = (a^2 + b^2 + c^2)\sigma^2 = \frac{3}{16}\sigma^2$$

and

$$\text{E}[L] = (a + b + c)\mu = \frac{3}{4}\mu,$$

so

$$\text{mse}[L] = \frac{3}{16}\sigma^2 + \left(\frac{3}{4}\mu - \mu\right)^2 = \frac{3}{16}\sigma^2 + \frac{1}{16}\mu^2.$$

This will be less than the mean square error of the unbiased estimator if $\dfrac{3}{16}\sigma^2 + \dfrac{1}{16}\mu^2 < \dfrac{1}{3}\sigma^2$ that is, if $\mu^2 < \dfrac{7}{3}\sigma^2$,

and greater otherwise. Because neither estimator has a smaller mse than the other for all values of the parameter μ, you cannot describe either as being more efficient than the other.

Consistency

As you have seen above, it is not necessary for an estimator to be unbiased. However, you would normally expect that an estimator based on a larger sample will come closer to, and be more certain of, identifying the true parameter value. You would hope that it would be more nearly correct on average, and that the spread about that average would decrease.

That is, if T_n is an estimator for τ based on a sample size of n, you would expect that as n gets larger,

- $\text{E}[T_n]$ would tend towards τ

- $\text{Var}[T_n]$ would tend towards 0.

An estimator with these properties is said to be *consistent*, and an equivalent definition is that the mean square error of the estimator tends to 0 as n gets larger.

EXAMPLE 7.5

Show that if $X_1, X_2, \ldots, X_n$ is an independent random sample from a distribution with mean μ and variance σ^2, the estimator

$$M_n = \frac{1}{n+1}(X_1 + X_2 + \ldots + X_n)$$

is a consistent estimator of the mean.

SOLUTION

$$E[M_n] = \frac{1}{n+1}\left(E[X_1] + E[X_2] + \ldots + E[X_n]\right) = \frac{1}{n+1}n\mu$$

$$= \left(1 - \frac{1}{n+1}\right)\mu$$

which tends towards μ as n gets larger, and

$$\text{Var}[M_n] = \left(\frac{1}{n+1}\right)^2\left(\text{Var}[X_1] + \text{Var}[X_2] + \ldots + \text{Var}[X_n]\right)$$

$$= \frac{n}{(n+1)^2}\sigma^2$$

which tends towards 0 as n gets larger.

Equivalently, the mean square error of M_n is

$$\text{mse}[M_n] = \text{Var}[M]_n + (E[M_n] - \mu)^2 = \frac{n}{(n+1)^2}\sigma^2 + \left(-\frac{\mu}{n+1}\right)^2$$

$$= \frac{n\sigma^2 + \mu^2}{(n+1)^2}$$

which tends towards 0 as n gets larger.

EXERCISE 7C

1 A coin is tossed n times and the number of heads, H, is counted.

If the probability of a head is p:

(i) Show that $\Pi_1 = \dfrac{H}{n}$ is an unbiased estimator of p, and find its standard error as a function of n and p.

(ii) Sketch a graph of the standard error as a function of p, for fixed n and show that the standard error obeys s.e. $\leqslant \dfrac{1}{2\sqrt{n}}$.

(iii) Roughly, what value of n is required to estimate p correct to 3 decimal places.

2 Errors are made by a typist at a rate of r per minute.

(i) The number of errors, N, he makes in t minutes is counted.

Assuming that N is a Poisson variable, write $E[N]$ and $\text{Var}[N]$ in terms of r and t.

Show that $R_1 = \dfrac{N}{t}$ is an unbiased estimator of r and find its standard error.

(ii) The number of minutes, T, he takes before making his nth mistake is counted.

Assuming that T is a random variable with density function

$$f(t) = \frac{r^n t^{n-1} e^{-rt}}{(n-1)!} \quad (0 \leqslant t < \infty)$$

write $E[T]$ and $\text{Var}[T]$ in terms of r and n. (You may find the integral

$$\int_0^\infty t^s e^{-kt}\ dt = \frac{s!}{k^{s+1}} \text{ helpful.})$$

Show that the estimator $R_2 = \dfrac{n}{T}$ of r is not unbiased and find its standard error.

Suggest an estimator similar to R_2, which is unbiased. Compare the mean square errors of R_2 and your estimator.

3 A bus comes every t minutes to my nearest stop. I use the bus rarely and my time of arrival at the stop when I do use it is independent of the bus timetable, so that the random variable W which records the time I have to wait for the bus on a particular occasion has a uniform distribution on $[0, t]$.

(i) Calculate the mean and variance of W.

I am going to estimate t by recording the value of W on the next six occasions on which I use the bus (call these values $W_1, W_2, \ldots, W_6$) and finding twice their mean, that is

$$T = \frac{W_1 + W_2 + \ldots + W_6}{3}$$

will be my estimate of the value of t.

(ii) Find $E[T]$ and $\text{Var}[T]$ (notice that $W_1, W_2, \ldots, W_6$ are independent), and hence show that T is unbiased and find its standard error.

The variable S is defined as the largest of my next six waiting times. The cumulative distribution function of S is

$$F(s) = P(S < s) = P(\text{all six } W_i < s)$$
$$= \{P(W < s)\}^6.$$

(iii) Calculate $F(s)$ and hence find $f(s)$, the p.d.f. of S.

(iv) Use the p.d.f. to calculate $E[S]$ and $\text{Var}[S]$.

Now consider the estimator $T_a = aS$.

(v) Find $E[T_a]$ and $\text{Var}[T_a]$ as functions of a and t.

(vi) For which value of a is T_a unbiased? What is the mean square error of T_a for this value of a?

(vii) Find an expression for the mean square error of T_a in terms of a, and hence determine the value of a which minimises the mean square error. What is the mean square error of T_a for this value of a?

(viii) Which estimator would you use?

4 When a random sample of size $(2r + 1)$ is taken from a random variable X, whose distribution has density function $f(x)$, and cumulative distribution function $F(x)$, the median of the sample has density function

$$m(x) = \frac{(2r + 1)!}{r! \, r!} (F(x))^r f(x)(1 - F(x))^r.$$

(i) Explain intuitively why this is the density function for the median.

Suppose X has a uniform distribution on $[0, a]$ and take $r = 1$, i.e. a sample of size 3.

(ii) Find the expectation and variance of:

(a) the median of the sample

(b) the mean of the sample.

(iii) Show that the median and the mean are both unbiased estimators of the mid-point of the distribution, and compare their standard errors.

(iv) Consider the variables $L_k = k \times \{\text{the largest of the three elements in the sample}\}$ as estimators of the mid-point of the distribution.

(a) Find expressions in terms of k for the mean and variance of L_k, and hence for its mean square error.

(b) Is there a value of k for which the estimator L_k is unbiased?

(c) Is there a value of k for which L_k is more efficient than the mean or the median?

5 A coin is tossed n times and the number of heads, H, is counted. If the probability of a head is p:

(i) Show that $\Pi_2 = \dfrac{H + 0.5}{n + 1}$ is not an unbiased estimate for p, and find an expression for the mean square error of Π_2 as a function of p. Is Π_2 consistent?

(ii) Compare Π_2 with the estimator Π_1 whose mse was determined in question 1. When is Π_2 preferable?

(iii) Even a biased coin is likely to have a value of p near to 0.5, where Π_2 has a smaller mse. Would Π_2 therefore be a better detector of biased coins?

(iv) Consider the estimator Π_3, which is just equal to 0.5, for all samples. Is this estimator consistent? Will this ever be the best estimator?

6 The time T at which an oil seal first fails is a random variable with density function

$$f(t) = \frac{t}{\lambda^2}e^{-\frac{t}{\lambda}} \quad (0 \leqslant t < \infty)$$

where λ is an unknown parameter.

(i) Find $E[T]$ and $\text{Var}[T]$.

A sample of n oil seals is tested until they fail. The times of failure are $T_1, T_2, \ldots, T_n$. Consider, as estimators of λ, the variables

$$S_c = c\sum_{i=1}^{n} T_i.$$

(ii) Determine the mean and variance of S_c and hence its bias and mean square error.

(iii) For what value of c is S_c unbiased?

(iv) For what value of c does S_c have minimum mean square error?

(v) Is S_c consistent for the values of c in **(iii)** and **(iv)**?

7 The independent random variables $X_1, X_2, \ldots, X_n$ represent a random sample from the $N(\mu, \sigma^2)$ distribution. The variance σ^2 is to be estimated.

Consider first the estimator $T = \frac{1}{2}(X_1 - X_2)^2$ which involves only X_1 and X_2.

(i) State the distribution of $X_1 - X_2$ and its mean and variance.

(ii) Use the result that $\text{Var}(X_1 - X_2) = E[(X_1 - X_2)^2] - \{E[X_1 - X_2]\}^2$ to deduce that $E[T] = \sigma^2$. State what this implies about T as an estimator of σ^2.

(iii) Using the result that $E[Z^4] = 3$ where $Z \sim N(0, 1)$, deduce further that $E[T^2] = 3\sigma^4$ and hence find $\text{Var}(T)$.

Now consider the estimator

$$S^2 = \frac{1}{n-1}\sum_{i=1}^{n}(X_i - \overline{X})^2$$

where $\overline{X}$ is the mean of $X_1, X_2, \ldots, X_n$. You are *given* the following results.

(1) The sampling distribution of S^2 is $\frac{\sigma^2}{n-1}\chi^2_{n-1}$.

(2) The mean of a χ^2_{n-1} random variable is $n - 1$.

(3) The variance of a χ^2_{n-1} random variable is $2(n - 1)$.

(You are *not* required to prove these results.)

(iv) Use these results to show that S^2 is an unbiased estimator of σ^2 and that its variance is $\frac{2\sigma^4}{n-1}$.

(v) Find $\frac{\text{Var}(T)}{\text{Var}(S^2)}$. Hence discuss the efficiency of S^2 relative to T as an estimator of σ^2.

(vi) Suggest possible circumstances when T might be a useful estimator of σ^2.

[MEI]

1 A variable X has a Cauchy distribution, that is, its density function is

$$f(x) = \frac{1}{\pi(1 + (x - \alpha)^2)}.$$

Two estimators based on a sample of size n have been suggested for α: the mean and the median.

Conduct a computer simulation to investigate the distributions of these two estimators: choose different sample sizes, and look at the spread of the distributions. Do you think both estimators are unbiased? Do you think both estimators are consistent? Which estimator is preferable?

The calculations are easier if you take $\alpha = 0$. In this case, if RAND is a variable with a uniform distribution on [0, 1] (which is provided by the BASIC language and by most spreadsheets), then

$$X = \tan(\pi(\text{RAND} - 0.5))$$

will have a Cauchy distribution.

2 An art historian, who knows that the signed and numbered copies of a Picasso print were numbered consecutively from 001 upward, has found five surviving copies, numbered 104, 078, 011, 083, 122.

Devise estimators for the number of copies made originally, and investigate their properties. Use the one you consider best to estimate the number of copies made from the data given.

1 If the form of the population distribution of a random variable is known or can be assumed, but the value of some parameter, τ of the distribution is unknown, then a statistic, T, calculated from a random sample of values of the variable can often be found which gives an *estimator* of that parameter. The value of the estimator from a particular sample is an *estimate* of the parameter.

2 The estimator is itself a random variable, and the probability, or probability density, with which each of its possible values arises in random samples from the population is called the *sampling distribution* of the estimator.

3 If the expected value of the estimator is equal to the population parameter, $E[T] = \tau$, the estimator is said to be *unbiased*.
Otherwise, if $E[T] \neq \tau$ the estimator is *biased* and the *bias* of T is $(E[T] - \tau)$.

4 The *standard error* of T is its standard deviation $se(T) = \sqrt{\text{Var}[T]}$.

5 The *mean square error* of T is defined as $mse(T) = E[(T - \tau)^2]$ which can also be calculated from $mse(T) = \text{Var}(T) + (\text{bias}(T))^2$.

6 One estimator of a parameter τ is more *efficient* than another if it has smaller mean square error for all values of τ.

7 Given a sample of n values $x_1, x_2, \ldots, x_n$ from a distribution with mean μ and variance σ^2, unbiased estimates of μ and σ^2 are given by:

$$\overline{x} = \frac{\sum\limits_{i=1}^{n} x_i}{n}$$

$$s^2 = \frac{\sum\limits_{i=1}^{n}(x_i - \overline{x})^2}{n-1} = \frac{\sum\limits_{i=1}^{n} x_i^2 - n \cdot \overline{x}^2}{n-1}.$$

8 Given a sample of n_1 values $x_1, x_2, \ldots, x_{n_1}$ from a distribution with mean μ_1, and variance σ^2, and a sample of n_2 values $y_1, y_2, \ldots, y_{n_2}$ from a distribution with a mean μ_2 (which may be different from μ_1) but the same variance σ^2, an unbiased estimate of σ^2 is given by

$$s^2 = \frac{\sum\limits_{i=1}^{n_1}(x_i - \overline{x})^2 + \sum\limits_{i=1}^{n_2}(y_i - \overline{y})^2}{(n_1 + n_2 - 2)} = \frac{(n_1 - 1)s_1^2 + (n_2 - 1)s_2^2}{(n_1 + n_2 - 2)}$$

where s_1^2 and s_2^2 are the unbiased estimates of σ^2 calculated from the two separate samples.

Answers

Chapter 1

Exercise 1A (Page 13)

1 $X^2 = 2.886$ (without Yates' correction)
  $= 1.852$ (with Yates' correction)
  $v = 1$
  Accept H_0 at 5% level or below: independent

2 $X^2 = 35.87$
  $v = 12$
  Reject H_0 at 5% level or above: association

3 $X^2 = 0.955$ (without Yates' correction)
  $= 0.602$ (with Yates' correction)
  $v = 1$
  Accept H_0 at 10% level or below: no association

4 $X^2 = 7.137$
  $v = 4$
  Accept H_0 at 10% level or below: independent

5 $X^2 = 10.38$ (without Yates' correction)
  $= 8.740$ (with Yates' correction)
  $v = 1$
  Reject H_0 at 1% level (c.v. $= 6.635$): related

6 $X^2 = 13.27$
  $v = 6$
  Reject H_0 at 2.5% level or above: not independent

7 $X^2 = 11.354$
  $v = 4$
  Reject H_0 at 5% level (c.v. $= 5.991$): association
  The cells with the largest values of $\dfrac{(o - e)^2}{e}$ are
  medium/induction and long/induction so medium and long
  service seem to be associated, respectively, with more than and
  fewer than expected employees with induction-only training.

8 $X^2 = 2.527$
  $v = 2$
  Accept H_0 at 5% level (c.v. $= 5.991$): no association

9 (i) $X^2 = 5.36$
  $v = 2$
  Reject H_0 at 10% level (c.v. $= 4.605$): association

(ii) Two degrees of freedom, since once the urban/none and
  urban/one values are fixed, all the other cell values follow
  from the row and column totals.

(iii) It appears that fewer rural residents than expected read
  more than one newspaper.

10 $X^2 = 22.18$
  $v = 9$
  Reject H_0 at 5% level (c.v. $= 16.92$): association
  Considering the values of $\dfrac{(o - e)^2}{e}$ for each cell, shows that
  rural areas seem to be associated with more reasonable and
  excellent and less poor or good air quality than expected.

11 (i) Expected frequencies: boys/cafeteria 11.44, boys/packed
  lunch 14.56; girls/cafeteria 10.56, girls/packed lunch 13.44

(ii) $X^2 = 4.121$, c.v. $= 3.841$ so reject H_0, seems to be
  association

(iii) $X^2 = 3.045$, c.v. $= 3.841$ so accept H_0, seems to be no
  association

(iv) If the result just achieves significance without Yates'
  correction and expected frequencies are not too large.

(v) Small sample; result close to critical value at 5% level;
  could quote significance level as between 5% and 10%.

12 H_0: no association, H_1: association
  $X^2 = 19.79$, c.v. $= 16.81$
  Reject H_0, seems to be association
  SE and Midlands have more short and fewer long lifespans
  than expected; 'rest' has fewer short and more long lifespans
  than expected.

Exercise 1B (Page 23)

1 (i) Estimate of $\lambda = 4.604$
  With 4 cells: $v = 2$, $X^2 = 0.375$ (depends on cell
  boundaries)
  So accept H_0 (c.v. $= 4.605$ even at 10% level): exponential
  is appropriate

(ii) Bus arrival times are unlikely to be independent because,
  for example, they run to a timetable, they are affected in
  the same way by traffic conditions.

2 (i) (a) $\bar{x} = 106.1$, $s = 16.37$
  With 7 cells: $v = 4$, $X^2 = 2.91$ (depends on cell
  boundaries)
  So accept H_0 (c.v. $= 7.779$ even at 10% level): normal
  is appropriate

(b) $t = 1.816$, $v = 9$
  So accept H_0 ($\mu = 100$) (c.v. $= 1.833$ even at 10% level)

(ii) $s = 39.5$
  With 7 cells: $v = 5$, $X^2 = 58.48$ (depends on cell
  boundaries)
  So reject H_0 (c.v. $= 16.75$ even at 0.5% level): N$(100, \sigma^2)$ is
  not appropriate

(iii) Accept, separately, that it is plausible that the distribution
  is normal and that it has mean 100; reject, with the
  combined sample, that it is plausible that the distribution
  is normal with mean 100.

3 $\bar{x} = 130.2$, $s = 39.89$

(i) With 5 cells: $v = 3$, $X^2 = 12.457$ (depends on cell
  boundaries)
  So reject H_0 at 1% level and above: given model is in
  appropriate

(ii) With 5 cells: $v = 2$, $X^2 = 1.024$ (depends on cell
  boundaries)
  So accept H_0 (c.v. $= 4.605$ even at 10% level):
  normal is appropriate

4 Sample mean 8.485

(i) Estimate λ as 0.2357

(ii) With 8 cells: $v = 6$, $X^2 = 7.696$ (depends on cell
  boundaries)
  So accept H_0 (c.v. $= 10.64$ even at 10% level): distribution
  is appropriate

(iii) Estimate λ as 0.3536

With 8 cells: $v = 6$, $X^2 = 2.810$ (depends on cell boundaries)

So accept H_0: the distribution is appropriate

(iv) The χ^2 test does not appear to be very powerful in discriminating between similarly shaped distributions.

5 With classes -29, 29–31, 31–32, 32–34, 34–

$v = 4$, $X^2 = 14.74$ (depends on cell boundaries)

So reject H_0 at 1% level and above: British population different

6 (i)

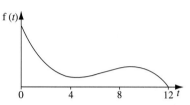

Most days tend to be either fairly sunny or overcast.

(iii) With 6 cells: $v = 5$, $X^2 = 10.29$ (depends on cell boundaries)

So accept H_0 at 5% level and below: model appropriate

(iv) The weather on consecutive days will not be independent.

7 $\bar{x} = 233.6$, $s = 27.45$

With 7 cells: $v = 4$, $X^2 = 11.42$ (depends on cell boundaries)

So H_0 (normally distributed) reject at 5% level and above

8 (i) $\dfrac{1}{\lambda}$

(iii) $\bar{x} = 4$

(iv) $v = 1$, $X^2 = 2.845$

Accept H_0 at 5% level: model is appropriate

9 (iii) $\bar{x} = 0.464$

(iv) $100 \displaystyle\int_0^{0.2} f(x)\, dx = 14.7$

(v) $v = 1$, $X^2 = 0.2731$

Accept H_0 at 5% level: model is appropriate

10 (iii) $\bar{x} = 10$

(iv) $0 \leqslant x < 5$: 23.61; $5 \leqslant x < 10$: 14.32; $10 \leqslant x < 15$: 8.68; $15 \leqslant x < \infty$: 13.39

(v) $X^2 = 10.31$, c.v. 5.991

(vi) Data appear uniformly distributed, not as the model predicts; no data in region 20 to ∞.

Chapter 2

Activity (Page 43)

1 If S is the least of $X_1, X_2, \ldots, X_n$, where $X_1, X_2, \ldots, X_n$ have distribution with the cumulative distribution function $F_X(x)$, then

$$F_S(x) = P(\text{least value in sample} \leqslant x)$$
$$= 1 - P(\text{least value in sample} > x)$$
$$= 1 - P(\text{all values in sample} > x)$$
$$= 1 - [1 - P(\text{a value of } X \leqslant x)]^n$$
$$= 1 - [1 - F_X(x)]^n$$

In Example 2.3, $f_S(x) = \frac{3}{8} e^{-3x/8}$ and $E[S] = \frac{8}{3}$; in Example 2.4 $E[S] = 0.151$; in Example 2.5

$$P(S = s) = \left(1 - \left(\frac{11}{12}\right)^3\right)\left(\left(\frac{11}{12}\right)^3\right)^{s-1} \text{ and } E[S] = 4.35.$$

Exercise 2A (Page 44)

1 (i) $E(P) = \pi$, $\text{Var}(P) = 0.7971$

(ii) $E(\bar{P}) = \pi$, $\text{Var}(\bar{P}) = \dfrac{0.7971}{n}$

(iii) n of the order of 10 million

2 mean 1.390, variance 0.0668

3 mean 0.6677, variance 0.1259

mean $\to \dfrac{2}{3}$ as $L \to \infty$

4 (i) $f(x) = \dfrac{(1 - u^2)(2 + 3u - u^3)}{8}$ where $u = \dfrac{(x - 20)}{3}$

mean of $L = 20.77$

(ii) mean of $H = 20.09$, variance $= 7.209$

(iii) mean of $M = 20.09$, variance $= 1.442$

5 (i) expectation 2.5, variance 1.25

(ii) $P(r) = \dfrac{3r^2 - 3r + 1}{64}$, expectation $\dfrac{55}{16}$, variance $\dfrac{143}{256}$

(iii) Distribution of sum

r	3	4	5	6	7
$P(r)$	$\frac{1}{64}$	$\frac{3}{64}$	$\frac{6}{64}$	$\frac{10}{64}$	$\frac{12}{64}$

r	8	9	10	11	12
$P(r)$	$\frac{12}{64}$	$\frac{10}{64}$	$\frac{6}{64}$	$\frac{3}{64}$	$\frac{1}{64}$

So mean of three scores has expectation 2.5, variance

$\dfrac{5}{12} = \dfrac{1.25}{3}$ as expected.

6 (i) $T = 60 \times \left(0.8 \times \dfrac{W}{70} + 0.2 \times \dfrac{C}{14}\right)$

(ii) $E(T) = 39.26$, $\text{Var}(T) = 25.98$

$P(\text{max 47 or more}) = 0.0777$

7 (i) $\dfrac{1}{1296}$

(ii) $P(X \leqslant 2) = \dfrac{1}{81}$; $P(X = 2) = P(X \leqslant 2) - P(X = 1) = \dfrac{5}{432}$

(iii) $\dfrac{65}{1296}$

(iv) $\dfrac{671}{1296}$; greater than $\dfrac{1}{2}$

8 (i) $f(x) = \dfrac{1}{4}$ $\quad (0 \leqslant x \leqslant 4)$

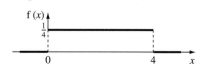

(ii) $F(x) = \begin{cases} \frac{1}{4}x & (0 \leqslant x \leqslant 4) \\ 0 & (x < 0) \\ 1 & (x > 4) \end{cases}$

(iii) p.d.f. of Y is $f(y) = \frac{1}{2}y$ $(0 \leqslant y \leqslant 2)$

9 (i) $N(9.9, (0.6)^2)$

  (ii) (a) 0.2523

  (b) 0.0334

  (c) 0.7143

  (iv) not independent

10 (ii) variance

  (iii) 0

  (v) $\mu_2 = \frac{3}{80}, \mu_3 = -\frac{1}{160}$. Not symmetrical

  (vi) $\mu_3 = 0$. Not symmetrical

Chapter 3

❓ (Page 53)

- Did you take a random sample of the appropriate population?
- Are the differences normally distributed? The differences should be continuous rather than discrete and you could also look at a stem-and-leaf diagram of your sample-differences: it should be unimodal and roughly symmetrical if the assumption of normality in the population is plausible.

Exercise 3A (Page 55)

1 (i) $t = 2.805, v = 13$, 1-tailed
Reject H_0 at 1% level or above: rats run more quickly when hungry

  (ii) Did the experimenters take a random sample of the relevant population of rats?
  It seems plausible that the differences in the rats times in the two conditions might be normally distributed.

  (iii) So that the effect of having learnt to run the maze on the second trial affects each condition in the same way.

2 (i) $t = 2.805, v = 13$, 1-tailed
Reject H_0 at 1% level or above: rating has improved

  (ii) No: the distribution of differences is discrete so not normal.

3 (i) $t = 1.528, v = 7$, 2-tailed
Accept H_0 at 10% level or below: no difference in ratings

  (ii) In question 2, the population is voters, the random variable whose mean is hypothesised to be zero is the change in voters' ratings; in question 3 the population is dates Feb to Oct and the random variable the difference in ratings of the two voters. In both cases, the random variable is assumed to be normally distributed in the population and the data constitute a random sample – in question 2 of voters, in question 3 of dates Feb to Oct.

4 (i) $t = 1.924, v = 16$, 1-tailed
Reject H_0 at 5% level or above: performance improves.

  (ii) Assumptions: random sample of population – you cannot tell how the subjects were selected; normally distributed differences in time between conditions – seems reasonable.

  (iii) The placebo test was done after the original test, there may be a learning effect improving the second attempt at the test.

5 (i) $t = 1.736, v = 6$, 1-tailed
Accept H_0 at 10% level or below: equivalent

  (ii) Assumptions: random sample of areas – could well have been; normally distributed differences in unemployment rates – not unreasonable.

  (iii) There may have been a change in economic conditions over the period of the test causing the unemployment rate to fall or rise nationally.

6 (i) $t = 0.6738, v = 8$, 2-tailed
Accept H_0 at 10% level or below: equivalent

  (ii) Assumptions: random sample of races – apparently not since all the sprints on just one afternoon are taken; normally distributed differences in timings – not unreasonable.

❓ (Page 61)

- Did you take two independent random samples of the appropriate population in the two conditions?
- Is the variable normally distributed in each condition? The differences should be continuous rather than discrete and you could also look at a stem-and-leaf diagram for each condition – they should be unimodal and roughly symmetrical if the assumption of normality in the population is plausible.
- Does the variable have the same variance in each condition? You could look at a stem-and-leaf diagram for each condition – they should have roughly the same spread if the assumption of equal variances in the population is plausible.

Exercise 3B (Page 64)

1 (i) $s = 2.646, t = 1.192, v = 17$, 1-tailed test
Accept H_0 even at 5% level: not heavier

  (ii) Assumptions: independent random samples on the two islands – it is hard to sample animal populations randomly – some individuals are more catchable than others; normally distributed weights on each island – seems reasonable; same variance of weights on each island – at first glance Daphne Major variance seems larger but samples are small.

2 (i) $s = 5.146, t = 0.304, v = 43$, 2-tailed test
Accept H_0 even at 10% level: same yield

  (ii)

```
              9|0|8 9
       0 4 2 4|1|2 4 3 4
   9 8 5 9 8 5|1|8 5 7 5 7 6 8 8
 3 2 1 0 1 1|2|0 0 4 0 0 3 0 3 3
       7 7 6|2|6 9
```

Assumptions: independent random samples of plots for the two varieties – you cannot tell how this selection was made; normally distributed yield with same variance in each variety – stem-and-leaf diagram suggests this is not unreasonable.

(iii) Select plots; plant some of each variety on half of each plot. An improvement since small average differences in yield are more likely to show up when the differences attributable to the plots are eliminated.

3 (i) $s = 1.568$, $t = 1.989$, $v = 40$, 1-tailed test

Accept H_0 at $2\frac{1}{2}\%$ level, reject at 5% level

(ii) Assumptions: independent random samples for the two groups – volunteer samples can be biased; normally distributed ratings with same variance in each group – probably not a reasonable assumption with data taking only a small number of discrete values.

(iii) Difficult because it is hard to pay two different rewards to the same person without alerting them to the point of the experiment.

4 (i) $s = 5.675$, $t = 2.085$, $v = 32$, 1-tailed test
Critical value at 5% level is less than 1.697
Assumptions: independent random samples of introverts and extroverts – you cannot tell how this selection was made; normally distributed heights with same variance in each group – nothing in data to suggest this is unreasonable.

(ii) No: causal conclusion is not justified – a confounding factor (e.g. features of childhood nurture) may produce both taller and more extrovert people.

5 (i) $s = 25.27$, $t = 1.264$, $v = 63$, 1-tailed test
Critical value at 5% level is more than 1.660
Assumptions: independent random samples for the two groups – hard to tell, but seems unlikely; normally distributed incomes with same variance in each group – not reasonable, as income distribution is likely to be skewed and the variance to be different in the two groups.

(ii) Those who have stayed on will have had fewer years in work. It would be better to investigate income eight years after completing education or income over time from age 24 to 30, say.

6 (i) $s = 3.453$, $t = 1.792$, $v = 23$, 1-tailed test

Accept H_0 at $2\frac{1}{2}\%$ level, reject at 5% level

(ii) Assumptions: independent random samples for the two floors – 'last year' is not a random sample of person-days and not independent for the two floors; normally distributed numbers of days absence with same variance on each floor – probably not a reasonable assumption with data taking only a smaller number of discrete values.

Exercise 3C (Page 68)

1 $t = -1.166$, $v = 22$, 1-tailed
Accept H_0 even at 5% level: not less
Assumptions: independent random samples of males and females; normally distributed lengths with same variance in each group.

2 (i) $t = 2.35$, $v = 11$, 1-tailed

Reject H_0 at $1\frac{1}{2}\%$ level and above: less

(ii) Assumptions: random sample of pairs; normally distributed differences between male and female lengths in mantis pairs.

(iii) Paired test is better able to discriminate difference due to sex in mantis size when other effects (e.g. age, diet, habitat) which affect male and female lengths similarly are eliminated by pairing.

3 $t = 1.005$, $v = 22$, 1-tailed
Accept H_0 even at 5% level: not more
Assumptions: independent random samples of patients for treatment or non-treatment; normally distributed post-operative stays with the same variance in each group.

4 (i) $t = 0.953$, $v = 6$, 2-tailed
Accept H_0 even at 10% level: 2% rise

(ii) Assumptions: random sample of economic conditions with crude oil price rising by 60% – you can obviously choose a random sample of countries, but only at one particular date, and this is not really the same thing; normally distributed changes in inflation rate.

(iii) No: the test is only of the model's prediction for this one event in the economic circumstances of the time – for instance, a second rise might find governments having a better idea how to respond to such a shock.

5 $t = 0.115$, $v = 9$, 1-tailed
Accept H_0 even at 5% level: 10.3 less
Assumptions: random sample of dyslexic adults; normally distributed differences in real and nonsense words learnt.

6 (i) $t = -1.243$, $v = 67$, 2-tailed
Reject H_0 even at 1% level: not 5.1 less

(ii) Assumptions: independent random samples of dyslexic and non-dyslexic adults; normally distributed numbers of nonsense words recalled with the same variance in each group. In this question assumptions are about two groups and the actual numbers of words they recall in one condition; in question 5, about one group and the difference in words recalled in two conditions.

(iii) The hypotheses are related. If D_r, D_n, N_r, N_n are the mean numbers of words recalled, respectively, by dyslexic adults recalling real words, dyslexic adults recalling nonsense words, non-dyslexic adults recalling real words and non-dyslexic adults recalling nonsense words, then the known information is that $N_r - N_n = 10.3$ and $N_r - D_r = 5.1$. Thus if the null hypothesis of question 5 is true, so that $D_r - D_n = 10.3$, this implies that $(N_r - N_n) - (D_r - D_n) = 0$ or $N_n - D_n = N_r - D_r = 5.1$ which is the null hypothesis of question 6.

Chapter 4

Exercise 4A (Page 77)

1 (i) $\bar{x} = 23.182$, $s = 4.708$, $v = 10$, $\tau = 2.228$
upper limit 26.34

(ii) (a) normally distributed
(b) samples chosen at random

2 (i) Interval is $\bar{x} - 2.821\dfrac{s}{\sqrt{10}} < \mu < \bar{x} + 1.833\dfrac{s}{\sqrt{10}}$, where $\bar{x}$ is the sample mean, s^2 is the unbiased sample estimate of the population variance and 1.833 and 2.821 are, respectively, the 5% and the 1% one-tailed critical values for the t-distribution with 9 degrees of freedom.

(ii) $\bar{x} = 0.4421$, $s = 0.03694$, $v = 9$
upper limit $(\tau = 1.833) = 0.4635$
lower limit $(\tau = 2.821) = 0.4091$

3 $\bar{x} = 1596$, $s = 339.5$, $v = 16$, $\tau = 2.587$ (interpolation)
lower limit 1383

Exercise 4B (Page 82)

1 (i) $\bar{x} = 23.182$, $s = 4.708$, $v = 10$, $\tau = 2.228$
upper limit 26.34

(ii) Assumptions: independent random samples of hen and duck eggs; normally distributed masses with the same variance for each type of egg.

2 (i) mean difference $= 2.632$, $s = 7.783$, $v = 18$,
$\tau = 1.736$ (interpolation)
interval 2.632 ± 3.255

(ii) Assumptions: random sample of pairs of brothers, born two years apart; normally distributed differences in salary.

3 mean difference $= 13.615$, $s = 10.31$, $v = 12$, $\tau = 2.681$
interval 13.615 ± 7.665

4 mean difference $= 17.111$, $s = 12.186$, $v = 14$, $\tau = 1.761$
upper limit 27.93

5 mean difference $= 241.48$, $s = 136.13$, $v = 31$,
$\tau = 1.696$ (interpolation)
interval 241.48 ± 81.31

Chapter 5

Exercise 5A (Page 103)

1 (i) $z = 1.5997$, 1-tailed test
Accept H_0 at 5% level, reject at 10% level

(ii) Assumptions: independent random samples from the appropriate populations of clover; appropriate to use the central limit theorem to justify a normal approximation to the distributions of sample means; variances of populations well approximated by their sample estimators.

2 $0.1852 \pm 0.0691\tau$; $\tau = 1.96$ for 95% interval
Assumptions: random sample of birds; appropriate to use the central limit theorem to justify a normal approximation to the distribution of sample means; variance of population well approximated by its sample estimator.

3 $z = 1.2869$, 1-tailed test
Accept H_0 at 5% level, reject at 10% level
Assumptions: independent random samples of waiting times; appropriate to use the central limit theorem to justify a normal approximation to the distributions of sample means; variances of populations well approximated by their sample estimators.

4 $z = 1.819$, 2-tailed test
Reject H_0 at 5% level, accept at 2% level

5 (i) $z = 2.379$, 2-tailed test
Reject H_0 at 2% level, accept at 1%

(ii) Assumptions: random sample of rate-differences; appropriate to use the central limit theorem to justify a normal approximation to the distribution of sample mean differences; variance of differences well approximated by its sample estimator.

The data could be seen as a random sample of targeting errors made at random by the Bank in their systematic attempt to shadow the Lombard rate, but it is hard to see how the values on 40 successive days could be seen as a random sample – each day's errors are likely to be correlated with the previous day's level of success.

6 $251.67 \pm 29.82\tau$
$\tau = 1.96$ for 95% interval

7 $z = 1.049$
Accept H_0 even at 10% level
Assumptions: independent random samples of boxes from the two machines; appropriate to use the central limit theorem to justify a normal approximation to the distributions of sample mean contents; variances of contents well approximated by their sample estimators.

8 (i) (a) $z = 1.6986$, 1-tailed test
Accept H_0 even at 2% level
(b) $z = 2.3034$, 1-tailed test
Reject H_0 even at 2% level

(ii) Either the variances in the two populations are, approximately, the same in which case this is an erratic pair of samples (since the sample variances differ) possibly leading to a type II error in conclusion **(a)**; or the variances are significantly different in the two populations and conclusion **(a)** is using an incorrect assumption and is hence unreliable.

9 (i) $z = 7.9847$, Reject H_0
Assumptions: independent random samples of subjects; appropriate to use the central limit theorem to justify a normal approximation to the distributions of mean numbers of words recalled; variances of numbers of words recalled well approximated by their sample estimators.

(ii) Prepare two lists of words A and B: select a single set of subjects, divided into four equal subsets and have each subset learn and recall one of: list A randomly followed by list B alphabetically, list B randomly followed by list A alphabetically, list A alphabetically followed by list B randomly, list B alphabetically followed by list A randomly.

10 (i) (a) $z = 2.934$, 1-tailed test
Reject H_0 at 1% level
(b) $z = 1.257$, 1-tailed test
Accept H_0 at 10% level

(ii) Interval is $14 \pm 4.77\tau$; $\tau = 1.96$ for 95% interval

11 (i) Lower bound 422.5
(ii) Interval 366 ± 5.15
Correct interpretation: this procedure leads to an interval including the true mean with probability 0.99.

12 (i) $\tau = 5.061$, $v = 9$, 2-tailed test: critical value 2.262
Reject H_0
(ii) $\tau = 1.974$, $v = 18$, 2-tailed test: critical value 2.104
Accept H_0
(iii) The first is a more discriminating analysis – a type II error is less likely because the paired design eliminates variation due to differences in resistance amongst components which might swamp the effect of the adjustment.

13 (i) 20.18 ± 1.10; 2-tailed test: critical value 2.262
(ii) $z = 2.439$, 2-tailed test: critical value 1.96
(iii) Use a Wilcoxon rank sum test – see page 129.

14 (i) H_0: Mean performance difference $= 0$.

H_1: Mean performance difference $\neq 0$.

(ii) Over the population of jobs, performance differences are normally distributed.

(iii) $t = 1.090$, $v = 7$, 2-tailed test: critical value 3.499

Accept H_0

(iv) 3.25 ± 7.05

15 (i) H_0: Mean arrivals at the two stations are equal.

H_1: Mean arrivals at the two stations are not equal.

(ii) Over the population of days, arrival numbers are independently and normally distributed at each station.

(iii) The variance of the arrival numbers is the same for each station.

(iv) $t = 2.507$, $v = 18$, 2-tailed test: critical value 2.104

Reject H_0

(v) Use a normal test (see pages 94–97)

16 (i) Filling volumes before and after overhaul are normally distributed with a common variance.

(ii) H_0: Mean volumes delivered before and after overhaul are equal.

H_1: Mean volumes delivered before and after overhaul are not equal.

(iii) $t = 0.9424$, $v = 15$, 2-tailed test: critical value 2.131

Reject H_0

(iv) No reason why an overhaul should not affect the variance of the volume delivered, so common variance assumption doubtful.

17 (ii) $E[T] = 32$, $Var[T] = 64$

(iii) The central limit theorem states that distributions of sums of many identical random variables are normal even if the underlying distribution is not: $n = 8$ is not large, but the underlying distribution is continuous and unimodal so you can expect a reasonable approximation.

(iv) $T \approx N(8\theta, 16\theta)$

(v) Other solution represents the upper limit of a one-sided 95% confidence interval.

18 (i) A single sample of tasks has been taken and the times taken by the two employees are not therefore independent samples as required by an unpaired design.

(ii) Differences in times on population of tasks is normally distributed.

$t = 2.274$, $v = 9$, 2-tailed test: critical value 2.262

Reject H_0

Confidence interval: 2.6 ± 2.1

(iii) Wilcoxon paired sample test – see pages 124–126.

19 $\bar{d} = 1.8$, $s^2 = 2.527$, $t = 2.996$

c.v. $= 3.143$ so accept H_0: treatment seems ineffective

Assumption: differences normally distributed.

Confidence interval: $(0.633, \infty)$

Interpretation: intervals determined in this way include the true mean difference in 95% of samples.

20 Means: 6.08, 5.02; $s^2 = 1.746$

$t = 1.8192$, c.v. $= 1.731$ so reject H_0: seems to be improvement

Assumptions: normal populations with the same variance.

Confidence interval: $(-0.16, 2.28)$

Interpretation: intervals determined in this way include the true difference of means in 95% of samples.

21 (i) $\bar{d} = 1.3$, $s^2 = 89.34$, $t = 0.4349$

c.v. $= 2.262$ so accept H_0: seems mean price the same.

(ii) $(-4.18, 6.78)$

(iii) differences normally distributed

(iv) Paired test compares the prices of the same item in both supermarkets and so eliminates variation, common to both supermarkets, in price between items.

22 (i) H_0: mean difference $= 0$, H_1: mean difference $\neq 0$

$\bar{d} = 1.8$, $s^2 = 2.277$, $t = 3.374$

c.v. $= 3.499$ so accept H_0: seems mean result the same.

(ii) $(0.538, 3.062)$

(iii) Dividing the sample eliminates possible variation between samples; sub-samples are assigned randomly in case the method of division is biased.

Chapter 6

Exercise 6A (Page 123)

1 $W = 19.5$, critical value $= 10$

Accept H_0: no difference

2 $W = 166$, critical value $= 175$

Reject H_0: score better

She must have made the claim before she saw that year's scores were better than the borough average.

3 $W = 714$, critical value $= 767$ (normal approximation)

Reject H_0: watches fast

4 $W = 15$, critical value $= 19$

Reject H_0: not correct

5 (i) $W = 16$, critical value $= 13$

Accept H_0: not greater

The percentage of moisture in samples of grain is symmetrically distributed about its median level.

(ii) $t = 1.933$, critical value $= 1.812$

Reject H_0: greater

The percentage of moisture in samples of grain is normally distributed.

(iii) The half of the data which is between the upper and lower quartiles is closely grouped around 3.4, there are long upper and lower quartile tails and the data is positively skewed. It is this odd sample distribution which produces a significant t-statistic and an insignificant Wilcoxon statistic. It could be an erratic sample (only 11 items) or, if it is representative of the population, both normal distribution and symmetry look unreasonable as assumptions.

6 There are 25 sets of ranks with sums less than or equal to 8 and

$\dfrac{25}{512} = 0.0488 < 0.05$, but 33 sets of ranks with sums less than

or equal to 9 and $\dfrac{33}{512} = 0.0645 > 0.05$.

7 (i) The distribution is likely to be skewed.

(ii) $T = 6$

(iii) (a) $T \leqslant 8$ **(b)** $T \leqslant 3$

(iv) There is some but not strong evidence against claim.

(v) $p = 0.0324$; good agreement

Exercise 6B (Page 137)

1 $W = 66.5$, critical value $= 65$
Accept H_0: no change

2 $W = 19.5$, critical value $= 21$
Reject H_0: improved

3 (i) $T = 100$, critical value $= 73$
Accept H_0: no difference

   (ii) Taking all the leavers from one school who went to university does not give a random sample of graduates.

4 $T = 490$, critical value $= 467$
Accept H_0: no difference
A stem-and-leaf diagram of the data suggests that the men's data are much more strongly bi-modal than the women's so the distributions differ not only in their location but also in their shape.

5 $T = 18$, critical value $= 19$
Reject H_0: harder

7 (i) (a) Both histograms show approximate normality.
      (b) The histograms are approximately the same shape.

   (ii) $T = 9$, critical value $= 10$
Reject H_0: different

8 (i) (b) $\{1,2,3,4\}, \{1,2,3,5\}, \{1,2,3,6\}, \{1,2,4,5\}$
      (c) $^9C_4 = 126$
      (e) 0.0331

   (ii) $T = 15$, critical value $= 8$
Accept H_0

9 (i) $W = 35$
   (ii) $W = 39$
   (iii) rejection
   (iv) Reject H_0

Calculate $\dfrac{\bar{x}_A - \bar{x}_B}{s\sqrt{\frac{1}{7} + \frac{1}{7}}}$, where $\bar{x}_A, \bar{x}_B$ are the sample means and s^2 is

the unbiased pooled-sample estimate of the population variance; compare this with the one-tailed 5% critical value for the t-distribution with 12 degrees of freedom.

10 (ii) $\dfrac{1}{2}m(m+1) + mn$
   (iii) $W = 52$, critical value $= 53$
Reject H_0

11 (i) $T = 3$, critical value $= 3$
Reject H_0
   (iii) 4.24%

12 (i) $T = 9$, c.v. $= 10$
Reject H_0: word processors differ
   (ii) $p = 0.0666$

13 (i) min 10, max 26
   (ii) 70
   (iv) $\dfrac{2}{70} \approx 0.02857$
   (v) $\dfrac{4}{70} \approx 0.05714$
   (vi) $p = 0.0606$
   (vii) $T = 11$; significant at 6% level

Chapter 7

❓ (Page 144)

See text that follows.

❓ (Page 147)

See text that follows.

Exercise 7A (Page 152)

1 (ii) $\dfrac{n}{n-1}V$ is unbiased

3 (i) (a) $L: \mathrm{P}(3) = \frac{1}{20}, \mathrm{P}(4) = \frac{3}{20}, \mathrm{P}(5) = \frac{6}{20}, \mathrm{P}(6) = \frac{10}{20}$

        $M: \mathrm{P}(2) = \frac{4}{20}, \mathrm{P}(3) = \frac{6}{20}, \mathrm{P}(4) = \frac{6}{20}, \mathrm{P}(5) = \frac{4}{20}$

      (b) $\mathrm{E}[L] = \frac{21}{4}, \mathrm{E}[M] = \frac{7}{2}$

      (c) $\mathrm{E}[2M - 1] = 6, \ \mathrm{E}[L] \neq 6$

   (ii) (a) $L: \mathrm{P}(1) = \frac{1}{56}, \mathrm{P}(2) = \frac{3}{56}, \mathrm{P}(3) = \frac{6}{56}, \mathrm{P}(4) = \frac{10}{56},$
           $\mathrm{P}(5) = \frac{15}{56}, \mathrm{P}(6) = \frac{21}{56}$

         $M: \mathrm{P}(1) = \frac{6}{56}, \mathrm{P}(2) = \frac{10}{56}, \mathrm{P}(3) = \frac{12}{56}, \mathrm{P}(4) = \frac{12}{56},$
           $\mathrm{P}(5) = \frac{10}{56}, \mathrm{P}(6) = \frac{6}{56}$

      (b) $\mathrm{E}[L] = \frac{119}{24}, \mathrm{E}[M] = \frac{7}{2}$

      (c) $\mathrm{E}[2M - 1] = 6, \ \mathrm{E}[L] \neq 6$

4 (i) $\mathrm{P}\left(\Pi = \frac{1}{3}\right) = 9p^2; \ \mathrm{P}\left(\Pi = \frac{1}{6}\right) = 6p(1-3p);$

        $\mathrm{P}(\Pi = 0) = (1-3p)^2$

   (ii) $9p^2 \times \dfrac{1}{3} + 6p(1-3p) \times \dfrac{1}{6} + (1-3p)^2 \times 0 = p$

5 (iv)

The estimate is biased for all values of p except 0 and 1.

   (v) If $p = 0$ then the experiment will never end so no value of N will be determined; if $p = 1$ then $N = 1, \Pi = 1$ and so Π is unbiased.

8 (iii) T_3 has a smaller variance than the other two estimators, so is less likely to give an estimate far from the true value of μ – see the work on mean square error later in this chapter.

9 (i) $\mathrm{E}[T] = k(n-1)\sigma^2, \ \mathrm{Var}[T] = 4k^2(n-1)^2\sigma^4$

10 (iii) $Z = 3$ or $5, \quad \mathrm{E}[Z] = 4, \quad \mathrm{Var}[Z] = 1$
   (iv) Y is biased, but has a small variance; Z is unbiased but has a larger variance for $n = 4$ or 5.

11 (iv) $E[Y] = \dfrac{n}{n+1}\theta,\quad Var[Y] = \dfrac{n\theta^2}{(n+2)(n+1)^2}$

(v) $Var[Z] = \dfrac{\theta^2}{n(n+2)}$

(vi) Both unbiased, but Z seems preferable as it has a smaller variance than $2\overline{X}$ (for $n > 1$).

12 (ii) $Var[T] = c_1^2\sigma_1^2 + (1-c_1)^2\sigma_2^2$

Minimum when $c_1 = \dfrac{\sigma_2^2}{\sigma_1^2 + \sigma_2^2}$, $c_2 = \dfrac{\sigma_1^2}{\sigma_1^2 + \sigma_2^2}$,

$Var[T] = \dfrac{2\sigma_1^2\sigma_2^2}{\sigma_1^2 + \sigma_2^2}$

(iii) T is a minimum variance unbiased estimator – it estimates correctly on average and is rarely far from the true mean – but it cannot be determined without knowing the variances of X_1 and X_2 which is unlikely in practice if μ is unknown.

13 (iv) $k = \dfrac{1}{n-1}$

14 Y has mean θ, variance θ, $E[Y^2] = \theta + \theta^2$. $\overline{Y}$ has variance $\dfrac{\theta}{n}$.

$\dfrac{1}{n}\sum Y_i^2$ unbiased estimator for $\theta + \theta^2$.

$a = 4, b = 2$

15 (iv) $E\left[\dfrac{1}{X}\right] = -\dfrac{p\ln p}{q}$ so biased

(v) $E[X^2] = \dfrac{2}{p^2} - \dfrac{1}{p}$; $a = \dfrac{1}{2}, b = -\dfrac{1}{2}$

Exercise 7B (Page 162)

1 (i) (a) 13.79

(b) 145.5

(ii) (a) 7.346

(b) 19.60

(c) 20.60

2 0.003 916

3 First 5, second 11

4 (ii) $s^2 = \dfrac{(n_1-1)s_x^2 + (n_2-1)s_y^2 + \dfrac{n_1 n_2}{n_1+n_2}(\overline{y}-\overline{x})^2}{n_1+n_2-1}$

(iii) Here, both samples are drawn from a single population, so a pooled estimate of the mean is made; in the text it was assumed that the samples were drawn from two populations and that only the variance was a common value to be estimated, so separate estimates of the mean were made.

5 (ii) $n\sigma^2$

(iii) $Var[\overline{X}] = \dfrac{\sigma^2}{n}$ so $E[n(\overline{X}-\mu)^2] = \sigma^2$

(v) 95% normal confidence interval has bounds:

mean $\pm 1.96 \times \sqrt{Var} = \sigma^2 \pm 1.96 \times \sqrt{\dfrac{2\sigma^4}{n-1}}$ and if n is large s can be used to approximate σ.

6 (iii) $E[X^2] = v(v+1) + 2\theta v + \theta^2$

(v) $E[\overline{X}] = $ mean of distribution $= \theta + v$;
$E[S^2] = $ variance of distribution $= v$, so S^2 is an unbiased estimator of v and $\overline{X} - S^2$ is an unbiased estimator of θ, since $E[\overline{X} - S^2] = E[\overline{X}] - E[S^2] = \theta$.

7 (iii) $E[X^2] = \sigma^2 + \mu^2$

(v) $E[\overline{X}^2] = \dfrac{\sigma^2}{n} + \mu^2$

(vii) $E[T] = \dfrac{n-1}{n}\sigma^2$

Exercise 7C (Page 172)

1 (i) $\sqrt{\dfrac{p(1-p)}{n}}$

(ii)

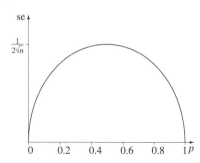

(iii) About a million

2 (i) $se = \sqrt{\dfrac{r}{t}}$

(ii) $E[T] = \dfrac{n}{r}$, $Var[T] = \dfrac{n}{r^2}$

$\dfrac{n-1}{n} R_2$ is unbiased with mse $\dfrac{r^2}{n-2}$;

mse $R_2 = \dfrac{(n+2)r^2}{(n-1)(n-2)}$

3 (i) $E[W] = \dfrac{t}{2}$, $Var[W] = \dfrac{t^2}{12}$

(ii) $E[T] = t$, $Var[T] = \dfrac{t^2}{18}$, $se(T) = \dfrac{t}{\sqrt{18}}$

(iii) $f(s) = \dfrac{6s^5}{t^6}$, $(0 \leqslant s \leqslant t)$

(iv) $E[S] = \dfrac{6t}{7}$, $Var[S] = \dfrac{3t^2}{196}$

(v) $E[T_a] = \dfrac{6at}{7}$, $Var[T_a] = \dfrac{3a^2t^2}{196}$

(vi) $a = \dfrac{7}{6}$, mse $= \dfrac{t^2}{48}$

(vii) mse $= \dfrac{t^2}{28}(21a^2 - 48a + 28)$ so $a = \dfrac{8}{7}$ gives minimum mse $= \dfrac{t^2}{49}$.

(viii) Very little difference in mse so might as well use unbiased estimator.

4 (i) Factorial expression gives number of ways of picking r elements of sample to be below and r elements to be above median value x; $(F(x))^r$ is the probability that the r elements picked are below x; $(1 - F(x))^r$ is the probability that the r elements picked are above x; and $f(x)$ is the density for the median value x itself.

(ii) (a) $E[\text{median}] = \dfrac{a}{2}$ $Var[\text{median}] = \dfrac{a^2}{20}$

(b) $E[\text{mean}] = \dfrac{a}{2}$ $Var[\text{mean}] = \dfrac{a^2}{36}$

(iv) (a) $E[L_k] = \dfrac{3ka}{4}$ $\text{Var}[L_k] = \dfrac{3k^2a^2}{80}$

$\quad\quad\quad \text{mse}(L_k) = \dfrac{12k^2 - 15k + 5}{20}a^2$

(b) L_k unbiased if $k = \dfrac{2}{3}$

(c) minimum mse at $k = \dfrac{5}{8}$ is $\dfrac{a^2}{64}$

5 (i) $\text{mse}(\Pi_2) = \dfrac{np(1-p) + (0.5-p)^2}{(n+1)^2}$

$\quad\quad$ Π_2 is consistent: mse tends to zero as n tends to infinity.

(ii) Π_2 has a smaller mse if p is near to 0.5, though only significantly so for small values of n.

(iii) No: smaller mse is not a criterion for an estimator to be discriminating; Π_2 will actually tend to disguise bias compared with Π_1.

(iv) Π_3 is not consistent (unless $p = 0.5$): mse $= (p - 0.5)^2$ for all n. It might be the most sensible estimator for small n if the coin is only slightly biased.

6 (i) $E[T] = 2\lambda$, $\text{Var}[T] = 2\lambda^2$

(ii) $E[S_c] = 2cn\lambda$, $\text{Var}[S_c] = 2c^2n\lambda^2$

$\quad\quad$ bias $(S_c) = (2cn - 1)\lambda$

$\quad\quad$ $\text{mse}(S_c) = \lambda^2\{2n(2n+1)c^2 - 4nc + 1\}$

(iii) unbiased when $c = \dfrac{1}{2n}$

(iv) minimum mse when $c = \dfrac{1}{1 + 2n}$

(v) consistent in both cases

7 (i) $N(0, 2\sigma^2)$

(ii) T is unbiased.

(iii) $\text{Var}[T] = 2\sigma^4$

(v) $\dfrac{\text{Var}[T]}{\text{Var}[S^2]} = n - 1$ so S^2 is more efficient than T and more so for larger n.

(vi) Gives an early rough estimate if sampling is slow or expensive.

Index